The Search for Life Continued
Planets Around Other Stars

W9-AOP-446

Barrie W. Jones

The Search for Life Continued

Planets Around Other Stars

Published in association with

 Springer

Praxis Publishing
Chichester, UK

Professor Barrie W. Jones
Physics and Astronomy Department
The Open University
Milton Keynes
UK

SPRINGER–PRAXIS BOOKS IN POPULAR ASTRONOMY
SUBJECT *ADVISORY EDITOR*: John Mason B.Sc., M.Sc., Ph.D.

ISBN 13: 978-0-387-76557-0 Springer Berlin Heidelberg New York

Springer is a part of Springer Science + Business Media (*springeronline.com*)

Library of Congress Control Number: 2008923617

Cover design: Jim Wilkie
Edited by: Laura Booth
Typesetting: BookEns Ltd, Royston, Herts., UK

Printed in Germany on acid-free paper

Contents

List of Illustrations

*To my wife Anne, to all others in my family
And in memory of my parents*

Preface

One of the most important astronomical discoveries of the twentieth century was of planets beyond our own Solar System. One of the most exciting prospects for the twenty-first century is the discovery of extraterrestrial life on such planets.

Like many other children, I was fascinated with astronomy. So much so, that in the 1950s, when I was in my teens, I built my own telescopes and used them to scrutinize the night sky. One object in particular grabbed my attention – the planet Mars. We need not look far for the reason – at that time it was widely thought that Mars could be inhabited, that some sort of life existed on its surface. Moreover, every two years or so Mars comes relatively close to the Earth, when it shines with a bright, steady light, red, like a drop of blood in the sky. At the time of its particularly close approach in 1956, the planet was intensively studied by the largest telescopes, as well as by me and many other amateur astronomers with our back garden telescopes!

But in spite of the widespread belief that there was life on Mars, the great majority of astronomers believed that this was probably no more than sparse vegetation, holding on in the cold, dry Martian deserts. It was thought unlikely that there were animals on Mars, and the possibility of intelligent Martians had long been confined to works of fiction. We now know that there is not even vegetation at the Martian surface. If there is life on Mars today, it is likely to be beneath the surface, and in the form of microbes. Nevertheless, this would still be a hugely exciting and important discovery, proving that the Earth is not a lonely abode of life.

It was in the 1980s that an even more promising habitable planetary body was identified in the Solar System. This is one of the satellites of the giant planet Jupiter. There are four large jovian satellites, and one of these, Europa, just might have living organisms in the oceans beneath its icy surface. One day, we will go there and find out.

It was in October 1995 that the first planet beyond the Solar System, orbiting a star rather like the Sun, was discovered. Other discoveries have followed, and now over 250 planets have been discovered, orbiting over 200 stars. The discovery of exoplanets, as they are called, soon prompted me to move much of my research and teaching into this area, and in particular, into investigations of whether there could be habitable planets in the known exoplanetary systems.

This book is my attempt to tell you where we are in the search for life in exoplanetary systems. I do hope that I will succeed in conveying to you the excitement of this quest. I have aimed the book at a wide readership, indeed at anyone who has an interest in answering the question, "Are we alone?" Therefore, I have assumed that you bring to this book no knowledge of astronomy. Mathematics, beyond a bit of simple arithmetic, is almost entirely absent. Embedded in the text are a few boxes. Some of these are asides, often biographical. A few outline some basic science that supports the main story.

I do hope that you enjoy this book and that it prepares you for what, I'm sure, will be stupendous discoveries in the not too distant future.

Acknowledgments

Dr Nick Sleep has provided assistance in several ways, including reading and commenting on a draft of the text.

Assistance with specific parts of the text have come from Professor Charles Cockell, Professor Charles Lineweaver, Dr Irene Ridge, Dr Jean Schneider, Dr Rachel Street, and Dr David Weldrake.

Illustrations have come from a variety of sources, nearly all of which are acknowledged in the captions. In a small number of cases the source has not been identified. We will be happy to make good in future editions any omissions brought to our attention.

Without the invitation from Clive Horwood of Praxis to write this book, it might never have been written.

1

Is there any life out there?

When you look up on a clear moonless night, many stars can be seen with the unaided eye, and many more are visible with binoculars or a telescope. Some of these stars are known to have planets, and the number is growing month by month. Do any of these planets harbor life?

This is a question of enormous public interest, one that I and my fellow astronomers are frequently asked. Yes! It is my belief that the Earth is not the only inhabited world in the cosmos, a belief based firmly on science. It is a belief

FIGURE 1.1 The starry sky – is there life out there? (© ESO, PR Photo 15a/04)

FIGURE 1.2 The Polish astronomer Nicolaus Copernicus (1473–1543), a portrait from his hometown Torun, in Poland, at the beginning of the sixteenth century.

shared by the great majority of scientists and by very large numbers of non-scientists. It seems inconceivable that in the vastness of the Universe the Earth is the only place where life exists. The Earth orbits the Sun – a star – and there is a truly huge number of stars visible in our telescopes. What an extraordinary conceit it would be to believe that there is life nowhere else.

Such a belief is reminiscent of the time when in Europe it was firmly thought that the Earth was special in being the center of the Universe – the geocentric system. Among the ancient Greeks there were a few who thought otherwise, and that the Sun was the center of the Universe – the heliocentric system. The first Greek astronomer to propose the heliocentric system was Aristarchus of Samos in around 270 BC, but this view did not start to spread until the sixteenth century, when the theory of heliocentrism was revived by the Polish astronomer Nicolaus Copernicus (Figure 1.2). Over the next two centuries heliocentrism became well established, on the basis of theory and observations.

However, well before heliocentrism became established, it had been proposed by the Italian philosopher Giordano Bruno (Figure 1.3), that even the Sun was not at the center of the Universe, but was merely one of the great number of stars. Observations in the eighteenth and nineteenth centuries showed that indeed this is the case. This was much too late for Bruno, who, in 1600, was burnt at the stake in Rome by the Catholic Church, largely for his belief in a plurality of worlds in an infinite Universe. It was ultimately discovered that the Sun is a fairly ordinary member of a vast assembly of about two hundred thousand million

FIGURE 1.3 The Italian philosopher Giordano Bruno (1548–1600).

stars called the Milky Way Galaxy (Figure 1.4). Worse still for human pride, in the twentieth century the Milky Way Galaxy was found to be just one of many hundreds of millions of galaxies. Thus was the Earth put in its place.

Like the belief in the geocentric system, throughout history there have been individuals, even groups, who have believed in the existence of life elsewhere. For example

> To consider the Earth as the only populated world in infinite space is as absurd as to assert that in an entire field of millet, only one grain will grow.

This is attributed to Metrodorus of Chios, a Greek philosopher, who lived in the fourth century BC. Another Greek philosopher, Epicurus, around 300 BC, said

> There are infinite worlds both like and unlike this world of ours . . . We must believe that in all worlds there are living creatures and planets and other things we see in this world.

Other prominent individuals throughout western history have held comparable beliefs, including Giordano Bruno. In his De L'infinito Universo E Mondi, published in 1584, he states

> There are countless suns and countless earths all rotating about their suns in exactly the same way as the seven planets of our system . . . The countless worlds in the Universe are no worse and no less inhabited than our Earth.

FIGURE 1.4 A galaxy that resembles the Milky Way Galaxy. The Sun is in the plane of the spiral arms, roughly half way from the center to the edge. (ESO, NGC1232)

From the seventeenth century onwards the belief that we are not alone has been more or less widespread in the western scientific community. This belief was fuelled by the move away from the geocentric system and by the invention of the telescope, which revealed the planets in the Solar System to be worlds, and not points of light.

In spite of this long-standing belief, the Earth remains the only place that we *know* to be inhabited. However, as you will see, this is because our searches for potential habitats beyond the Earth have barely begun. You will also see that, on good scientific grounds, we have great expectations of finding life elsewhere. It is thus beyond reasonable doubt that, just as the Earth has been moved from the center of the Universe, it will also be removed from being the only place where life is found. The failure to discover extraterrestrial life before the end of this century would be a surprise to the scientific community, and of enormous significance. What an exciting era we are entering.

In this book I concentrate on life *beyond* our Solar System. There are two main reasons for this. First, books on life in the Universe tend to concentrate on life *in* the Solar System, so this topic is very well covered elsewhere. Conversely, life beyond the Solar System is less well covered, yet this is at a time when we are beginning to discover planetary systems beyond our own – at the time of writing (late 2007) over 200 planetary systems beyond our own are known, a number that is growing by the month. In the next decade some of these exoplanets will be scrutinized to see if they are potential habitats, and indeed whether they are inhabited.

Second, intelligent life is highly unlikely to be found elsewhere in the Solar System – the search for extraterrestrial intelligence must concentrate its efforts further away.

But even though it is not the focus of this book, I cannot omit the Solar System entirely. After all, the Earth is the only place where we *know* there is life, and the Earth is one of the planets in the Solar System. Life here will show us what it is we are looking for out there, and will guide us toward the best search strategies. This might seem parochial – what about alien biologies? These are not entirely neglected, but life as we know it is where we must start.

Consequently, in Chapter 2 I present a brief description of the Solar System, and a brief account of its origin. This places planet Earth in context, and prepares the ground for our consideration of other planetary systems that might have life. It will also establish the planetary backdrop for Chapters 3–5, where I will outline the nature of life on Earth, its possible origin, its evolution, and its future.

It would be perverse to charge straight from the Earth to planets around other stars when we have such ripe candidates for extraterrestrial life in our "back garden". Therefore, Chapter 6 is devoted to a brief look at potential habitats elsewhere in the Solar System, particularly the planet Mars and the satellite of Jupiter called Europa.

We then take our leave of the Solar System, and consider where habitats might be found beyond it, the topic of Chapter 7. Chapters 8–10 describe how searches are being made for such habitats. Chapter 11 summarizes the results to date and Chapter 12 looks to future discoveries. Chapter 13 discusses how a potentially habitable planet can be investigated for life, including the frustrating outcome "case not proven" that could only be resolved by interstellar travel.

The search for extraterrestrial intelligence (SETI) is described in Chapter 14. This is now a topic that astronomers can write about without fear of sideways glances by other scientists. The broader public has long been more sympathetic to this search. The chapter will include a brief discussion of whether ETI has been detected, a question which often generates more heat than light. It is almost certain that aliens, intelligent or otherwise, have *not* been seen. Of course, this need not prevent us indulging in the fascinating speculation of the external appearance of aliens, and what they might be like internally. Chapter 15 presents some possibilities, based on scientific constraints, and includes life *not* as we know it.

But before you embark on your journey through these pages, think of this simple journey you can make yourself. Go out at night under a clear dark sky and look upwards. It is beyond reasonable doubt that a few of the stars that you can see with the unaided eye will each have at least one planet that bears life, of some sort.

2

The Solar System

The Earth is one of many bodies in the Solar System. In this chapter I will describe the Solar System and the bodies within it. This will not only introduce you to our home and our cosmic backyard, it will also introduce you to science that will be essential for later chapters.

2.1 ORBITS AND SIZES

Figure 2.1 shows a plan view of the orbits of the eight planets around our star, the Sun. The planets themselves, and the Sun, which is far larger than any planet, are too small to be shown on this scale. The scales are marked on the Figure. The upper scale is 150 million kilometers (1 kilometer (km) = 0.621 miles). This is very nearly the same as the average distance between the Earth and the Sun, which is called the astronomical unit, abbreviated to AU, and equal to 149.6 million km. The AU is nearly 12,000 times the diameter of the Earth – quite a distance.

The orbits of the planets are not quite circular – they are ellipses, which have the shape of a circle when it is viewed at an angle. The non-circular shape is not apparent on the scale of Figure 2.1. What *is* apparent, particularly for Mercury and Mars, is that the Sun is not quite at the center of the orbit. This is a consequence of the ellipticity. Figure 2.2 shows an orbit far more elliptical than any of the planetary orbits in the Solar System. On this orbit two quantities are shown that will be important in the chapters on planets beyond the Solar System. These are the size of the orbit as measured by its semimajor axis a ($2a$ is shown), and the non-circularity of the orbit as measured by its eccentricity e (e times a is shown).

The orbits are also not quite in the same plane. Mercury is the most inclined, but only at $7.0°$ with respect to the orbital plane of the Earth. This plane (called the ecliptic plane) is the reference plane in the Solar System.

Figure 2.3 shows the relative sizes of the Sun and the eight planets. You can see that the four closest to the Sun – Mercury, Venus, Earth, and Mars – are far smaller than Jupiter, Saturn, Uranus, and Neptune. These two groups are called, respectively, the terrestrial planets and (unsurprisingly) the giant planets. Note that the Sun is nearly 10 times the radius of the largest planet, Jupiter. Even so,

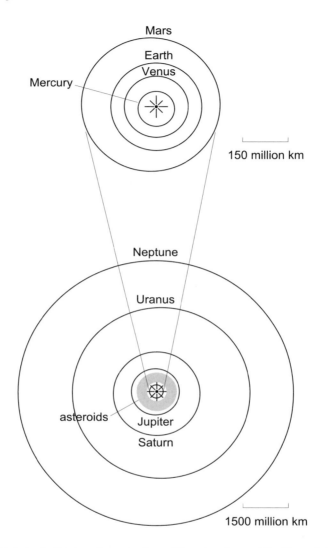

FIGURE 2.1 The Solar System – a face on view of the planetary orbits. Though the orbits are not quite in the same plane, this makes no difference to the view on the scale here.

the Sun's radius is still 215 times smaller than the radius of the Earth's orbit. The Sun and the planets are very nearly spherical. When comparing the sizes of two spheres remember that the volume ratio is far larger than the ratio of the radii. The volume is proportional to the cube of the radius. For example, if one sphere has twice the radius of another it has $2 \times 2 \times 2 = 8$ times the volume. Therefore the Sun is approximately $10 \times 10 \times 10 = 1,000$ times the volume of Jupiter.

Figure 2.4 shows images of the planets (and the Moon). This is to give you a

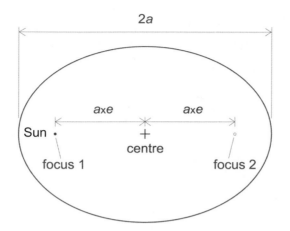

FIGURE 2.2 The orbit of the comet 21P Giacobini-Zinner, to show what is meant by the semimajor axis a of an orbit and its eccentricity e, which is 0.7057. If $e = 0$ the orbit is a circle centered on the Sun. The Sun is at one of the two points called the focuses of the ellipse. The other focus is empty.

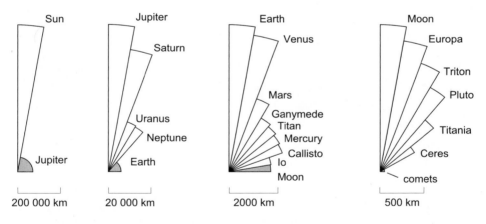

FIGURE 2.3 The Solar System – the Sun and its eight planets, showing relative sizes, and also the larger planetary satellites, the largest asteroid Ceres, and the dwarf planet Pluto.

visual impression of their appearance from a spacecraft. Comparison with Figure 2.3 shows that these images are *not* to scale.

As well as the Sun and the eight planets there are several other types of body in the Solar System, some of which are shown in Figure 2.3 (most of these are the large planetary satellites, of which, more shortly). You might be surprised that Pluto is not included among the planets. Until recently there were *nine* planets in the Solar System – the eight in Figure 2.4, plus Pluto. At the General Assembly of the International Astronomical Union, held in Prague in August 2006, Pluto was

FIGURE 2.4 The eight planets (not to scale). From top to bottom: Mercury, Venus, Earth, Mars, Jupiter, Saturn, Uranus, Neptune. The Moon is top right. (NASA/JPL-Caltech, PIA 03153)

demoted. One reason for this is its size. There is a real step down in size from Mercury (the smallest of the eight other planets) to Pluto (Figure 2.3). Pluto's radius, 1,153 km, is 47% that of Mercury, so its volume is only 11% that of Mercury. Another reason for Pluto's demotion is that beyond Pluto (which lies beyond Neptune) other bodies are being found that are comparable in size to, or even larger than, Pluto. One such is called Eris, which is even further away than Pluto, which is why it was only discovered in 2003.

Pluto, Eris, and a growing number of other bodies, are now called dwarf planets. The official name of the eight larger bodies is simply "planets". Future revisions to this naming are not out of the question.

Pluto and Eris are two members of a group of bodies beyond Neptune that constitute the Edgeworth–Kuiper belt, named after the Anglo–Irish astronomer Kenneth Essex Edgeworth (1880–1972) and the Dutch–American astronomer Gerard Peter Kuiper (1905–1973). Not all of the belt members are dwarf planets. To be a dwarf planet (or a planet) a body has to be large enough for its self-gravity to pull it into a close-to-spherical shape. Bodies too small for this to be the case are classified as small Solar System bodies.

Another population of small Solar System bodies is the asteroids. These orbit mainly between the orbits of Mars and Jupiter (Figure 2.1). The largest asteroid, Ceres ("series"), has a radius of 479 km, just large enough for its internal gravity to pull it into a shape close to spherical, and so Ceres is also classified as a dwarf planet, though it is still the largest asteroid! The asteroids are the main source of the meteorites – small bodies that arrive on the Earth's surface from space.

The Edgeworth–Kuiper belt is the source of many of the comets that visit the inner Solar System. The Oort cloud is the other major source. This cloud surrounds the Sun, but even its inner members lie at least 20 times further from the Sun than Neptune. Comets are tiny icy-rocky bodies, rarely larger than a few tens of kilometers across, but they develop spectacular tails when they enter the inner Solar System, formed from material driven off by solar radiation (Figure 2.5).

Last, but not least, there are the natural satellites ("moons"). All major planets (and some dwarf planets) have satellites, except Mercury and Venus. Mars has two tiny satellites. The Earth has just one, the Moon. This has a radius of 1,738 km. The radius of the Earth is 6,378 km, making the Moon larger compared to its planet than any other satellites of a planet. But the Moon is by no means the largest satellite. Among the many satellites of the giant planets two are larger than Mercury (Figure 2.3). These are Jupiter's largest satellite Ganymede and Saturn's largest satellite (by far) Titan. Also shown are the other large satellites. Callisto, Europa, and Io are satellites of Jupiter. Triton is by far the largest satellite of Neptune, and Titania, by a short head, is the largest satellite of Uranus.

In spite of their size, large satellites are not classified as planets, and somewhat smaller satellites are not classified as dwarf planets. This is because, by definition, a satellite orbits a planet, rather than the Sun.

Let's now look at a few Solar System bodies in more detail, particularly the Sun and the Earth, but also, briefly, the other bodies, including the giant planets.

FIGURE 2.5 The comet Hale-Bopp, which passed through the inner Solar System in 1997. (Dr Francisco Diego, University College London)

2.2   THE SUN

Without the Sun, the surface of the Earth, and of the other three terrestrial planets, would be at temperatures far below 0°C (32°F). Even though the internal heat sources of the Earth could keep all but the upper layers of the crust warm, the upper few hundred meters and the surface, except in volcanically active regions, would be far too cold for life, and the oceans almost everywhere would be frozen as hard as iron. Life at and near the surface would be impossible over nearly all of the Earth.

The surface of the Sun, called its photosphere, has a temperature of about 5500°C. This is called the Sun's effective temperature. It is this high temperature on a body with such a large surface area that makes the Sun luminous enough to warm the terrestrial planets. The Sun has warmed the planets throughout the 4,600 million years since the Sun and the planets were born. What could have made the Sun so luminous over such an enormous span of time? The answer lies deep inside the Sun.

The solar interior

The Sun has a radius about 100 times that of the Earth and therefore a volume about a million times greater. It is a fluid throughout. The Sun consists almost entirely of the lightest two chemical elements, hydrogen and helium. When the Sun was born, these elements accounted respectively for about 71% and 27–28% of the Sun's mass, leaving rather less than 2% for all the other 90 chemical elements. By contrast, the terrestrial planets, being rocky, consist almost entirely of these other elements.

Figure 2.6 shows a cut-away of the Sun. Temperatures increase rapidly with depth, and are everywhere high enough for the atoms to have lost their electrons, to become ionized, in a process called ionization. Within the core the temperatures are high enough for a process called thermonuclear fusion to occur – this is what defines the core. Nuclear fusion is the process in which two atomic nuclei are hurled together so fast that the nuclei fuse together to make a different nucleus. The higher the temperature the greater the speed of the nuclei, and the more often nuclear fusion occurs, hence the prefix "thermo". In the Sun's core, the overall effect of a series of thermonuclear fusion reactions is to convert four hydrogen nuclei into one helium nucleus – four protons have been joined to make a nucleus consisting of two protons and two neutrons. In this process a tiny amount of energy is released, mainly in the form of gamma rays (see below) and subatomic particles.

At temperatures above about 10 million°C the fusion of hydrogen to helium in the Sun's dense core happens so often that a lot of energy is released. The core of the Sun has always been hotter than this (it is currently about 14 million°C). Consequently, since its birth, the helium proportion in the Sun's core has increased at the expense of hydrogen.

Box 2.1   The structure of the atom and the chemical elements

The atom consists of a tiny nucleus made up of a certain number of electrically neutral particles called neutrons, and a certain number of positively charged particles called protons. The nucleus is surrounded by a swarm of tiny negatively charged particles called electrons. For an atom to be electrically neutral, the number of protons has to equal the number of electrons, otherwise it is called an ion. A proton and a neutron are each around 1,840 times the mass of an electron, so nearly all the mass of the atom is in the nucleus. By contrast, the volume of the atom is determined by the electrons, which swarm in a volume far larger than that of the nucleus. Atomic radii are so small that, placed side by side, a few million would be needed to span a millimeter.

The nucleus is typically about 100,000 times smaller than the orbits of the electron. To put this into a human scale, imagine that a nucleus had a diameter of 0.1 millimeters – a tiny pinhead. The electron orbits would then stretch out to about 100 meters.

The simplest atom is that of hydrogen, denoted by H. It consists of a nucleus of one proton surrounded by one electron. This makes it the least massive atom and the least massive nucleus of all. Next is deuterium, in which the nucleus consists of one proton and one neutron, so it is about twice the mass of hydrogen. Like hydrogen the deuterium nucleus is surrounded by one electron. The chemical properties of an atom are determined by the number of electrons in its neutral form, which is equal to the number of protons in the nucleus. The number of protons, called the atomic number, defines a chemical element. Thus, hydrogen and deuterium are the same chemical element, which is called hydrogen. Hydrogen and deuterium are called different isotopes of hydrogen.

The next element must have two protons in its nucleus. This is helium (He). The common isotope has two neutrons in the nucleus, making it four times as massive as H.

And thus we build up the chemical elements, with no gaps in the number of protons. As a final example consider carbon (C), which has six protons in its nucleus. The most common isotope by far has six neutrons, making 12 nuclear particles in all. This is called carbon-12. There are also isotopes with four, five, seven, eight, or nine neutrons – carbon-10, carbon-11, carbon-13, carbon-14, and carbon-15.

Not all isotopes are stable. Some decay into other isotopes. Such unstable isotopes are radioactive – they emit particles or radiation.

It is the core fusion that makes the Sun so luminous. The radiation from the photosphere has warmed the Earth's surface throughout its history, essential for the presence of life at the surface. The Sun's luminosity is not constant. On a timescale of roughly a million years (Myr) the variations are slight, but since the Sun was born 4,600 Myr ago its luminosity has increased from about 70% of its

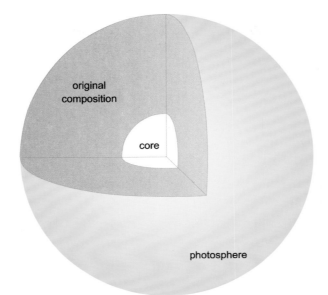

FIGURE 2.6 The Sun, showing the core where hydrogen fusion occurs.

present value. Its luminosity will continue to increase until in a few thousand Myr from now the Earth will have become too hot for life. Further in the future, about 6,000 Myr from now the hydrogen in the core will all have been fused to helium – this marks the end of what is called the main sequence lifetime of the Sun, which started at its birth. The Sun will then start on its transition to becoming a red giant star (Figure 2.7), and nowhere in the Solar System will be habitable.

Box 2.2   A cosmic clock tick

A million years is written 1 Myr, where "M" stands for "million" and "yr" stands for "year". It is a long time compared to a human life-span, but it is a convenient unit of time when we are considering the history of a planet or a star. You can think of 1 Myr as the tick of a cosmic clock, and so there have been 4,600 clock ticks since the Earth was born. The Myr unit of time will be used throughout this book.

The solar spectrum and electromagnetic radiation

To further lay a basis for later chapters, you need to know something about what is called the electromagnetic spectrum. The solar spectrum is a good way to introduce the essential ideas.

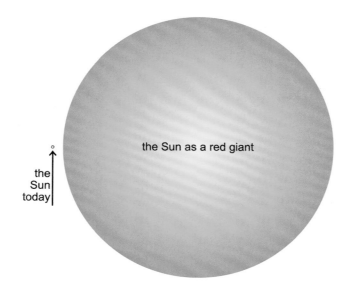

FIGURE 2.7 The Sun about 7,000 Myr from now, when it will be a fully fledged red giant, and the Solar System will be uninhabitable.

When you look at the Sun, our human eyes are almost blinded by the light. However, the luminosity of the Sun does not consist only of light, but also of infrared radiation and other forms of radiation too, notably ultraviolet radiation. All these forms of radiation are examples of what is called electromagnetic radiation. As its name suggests, it has a magnetic and an electric component. It would take us too far afield to go into this in much detail. The essential points are that electromagnetic radiation consists of waves, oscillating around zero, and travelling through space at 299,792 km per second – the speed of light. It is the wavelength of these waves that distinguishes the different forms (Figure 2.8).

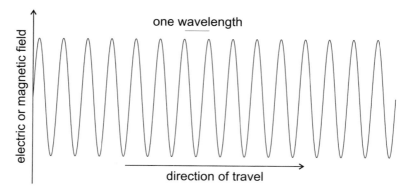

FIGURE 2.8 A wave, illustrating the concept of wavelength.

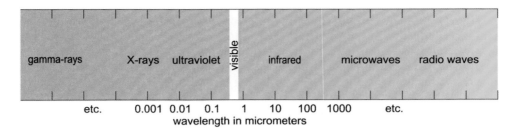

FIGURE 2.9 The electromagnetic spectrum. A micrometer is a millionth of a meter.

Figure 2.9 shows what is called the electromagnetic spectrum. The wavelength increases from the left, in leaps and bounds – from tickmark to tickmark the wavelength increase ten fold. You can see that visible radiation, light, covers a narrow range, with violet to the left and red to the right. Across the visible range the order is as in a rainbow – violet, blue, green, yellow, orange, and red (indigo seems to be a fiction to reach the mystical number seven – there's no convincing color between blue and violet). The wavelengths of light extend from about 0.40–0.70 millionths of a meter, though some people can see slightly beyond this range.

Infrared radiation has longer wavelengths than red light, and we can sense it with our skins. Ultraviolet radiation has shorter wavelengths than violet light, and our skins are sensitive to it too, but unfortunately with a delayed response that can result in sunburn.

The Sun's luminosity is made up mainly of visible and infrared radiation. This is shown in Figure 2.10, which is also called a spectrum, though unlike Figure 2.9 it shows the relative quantities of radiation at each wavelength. Note how rapidly the quantities fall off to left and right.

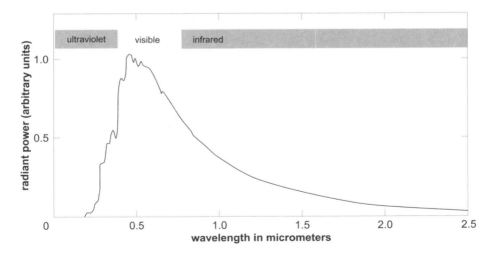

FIGURE 2.10 The solar spectrum. A micrometer is a millionth of a meter.

There's just one more thing you need to know about electromagnetic radiation, and this is the photon. I have said that electromagnetic radiation is a wave, and indeed it is. But it's a curious phenomenon. Light moves from place to place displaying wave-like properties, such as spreading around corners, rather like the surface waves on water. But when it interacts with matter it can behave like a hail of bullets called photons. A photon carries energy, the shorter the wavelength the greater the energy. A photon of ultraviolet radiation carries more energy than radiation at any visible wavelength, and it is this that makes ultraviolet radiation damaging to our skins, in spite of the rather small quantity in the solar spectrum (Figure 2.10). When considering the interaction of electromagnetic radiation with matter, the photon aspect is particularly pertinent.

2.3   THE EARTH

Before I turn to life on Earth, I need to present you with the planet's major features – the home is described before the inhabitants.

The Earth is very nearly spherical, slightly flattened by its rotation once per day. This rotation is around an axis through its center, and where this axis emerges through the surface we have the North and South Poles (Figure 2.11). The Earth has an average equatorial radius of 6,378 km, and a polar radius of 6,357 km. You can get a feel for the size of these radii by considering the shortest distance across the Earth's surface between London and New York – 5,570 km.

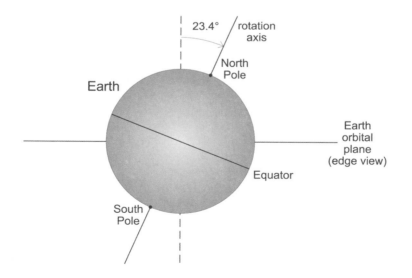

FIGURE 2.11 The Earth's rotation axis, tilted by 23.4 degrees from the perpendicular to the Earth's orbital plane.

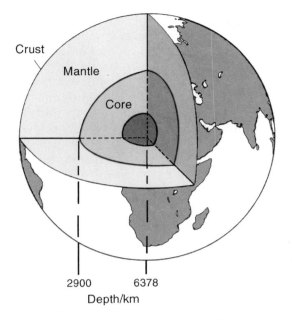

FIGURE 2.12 A cut-away of the Earth, showing the major components.

Internal structure and composition

Unlike the Sun, the Earth is almost entirely made of rocky materials and iron, which are made from chemical elements that constitute somewhat less than 2% of the Sun's mass. Figure 2.12 shows a cut-away of the Earth, revealing its major components. The core consists mainly of iron (Fe), liquid in a thick outer shell but with a solid inner core. At the center the temperature is estimated to be about 5200°C. The liquid shell is the source of the Earth's magnetic field.

Above the core lies the mantle, which is solid, and accounts for about 83% of the volume of the Earth. It consists almost entirely of rocky materials, notably silicates. These are compounds of one or more metallic elements, combined with the abundant elements oxygen (O) and silicon (Si). Common among the metallic elements are iron, magnesium, and calcium. The mantle is shown as one unit in Figure 2.12, which will suffice for our purposes, though there are divisions due to changes, predominantly with depth, in mineral crystal structure and to a minor extent in composition.

Above the mantle we come to the topmost layer, the crust. This also consists predominantly of silicates, though of substantially different composition from those that constitute the mantle. In particular, the crustal silicates are richer in the metals calcium and aluminium, and poorer in iron and magnesium. They have been derived from the mantle silicates, by partial or complete melting. There are, in fact, two types of crust, the oceanic crust, with a thickness varying from about 5 km to about 10 km, and the continental crust, thicker, from about

20 km to about 100 km. As their names suggest, they are found (predominantly) under the oceans and beneath the surfaces of the continents. However, it is a coincidence that the present volume of the oceans raises sea level approximately to where continental crust meets oceanic crust.

The two types of crust differ in composition, with continental crust even more depleted than oceanic crust in iron and magnesium. Continental crust, particularly in its upper reaches, is also enriched in silicon and oxygen. The production (and destruction) of crust is outlined briefly below, in *Plate tectonics –* a process intimately related to the evolution of life.

Though the continental crust is made of silicates, these are visible at the Earth's surface in unmodified forms as rocky outcrops only over a small proportion of the continental areas. This is because the initial silicates have been changed by a variety of processes, including physical weathering by water and wind, and chemical modification by water, atmospheric gases, and the biosphere. Some of the products have subsequently been modified by high pressures and elevated temperatures. We thus have landscapes dominated by soil, clays, sand and rocks such as limestones and sandstones. Only beneath such deposits do we meet relatively unmodified crust.

Plate tectonics

We turn now to the production and destruction of the Earth's crust. The Earth's surface is divided into a few dozen plates. These are in motion with respect to each other, being created at some boundaries, sliding past each other at other boundaries, and being destroyed at boundaries where one plate dives beneath its neighbor. Each plate consists of crust and the upper few tens of kilometers of the mantle. Each of these components is comparatively rigid and together constitute the lithosphere (Greek "lithos" = "stone"). The underlying mantle, although not liquid, is much less rigid, and this allows the plates to move with respect to each other.

Figure 2.13 shows a section through adjacent lithospheric plates. In the center of the Figure two plates carrying oceanic crust – oceanic plates – are moving apart, allowing fresh oceanic crust to form from partial melting of the mantle, which melts as it rises due to the decrease in pressure. This modifies its (silicate) composition. To the right there is destruction of the oceanic plate that is diving beneath the other oceanic plate – this is called a subduction zone.

To the left the subduction is carrying the oceanic plate beneath a plate bearing continental crust – a continental plate. The continental plate is less dense and therefore more buoyant than the oceanic plate, so continental plates never do the diving. Continental crust forms from partial melting of the mantle of the overriding plate, which again modifies the composition, followed by further modification as the melt rises through the existing continental crust.

You can see in Figure 2.13 that oceanic crust is created where two plates are moving apart, and that it is destroyed at subduction zones. This has kept the volume of oceanic crust roughly constant. The formation of *continental* crust is

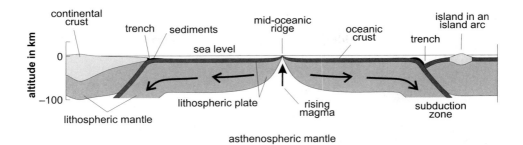

FIGURE 2.13 A section through the Earth's crust and upper mantle, showing some of the main features associated with plate tectonics.

almost entirely at subduction zones, and for the last 2,500 Myr or so out of the 4,600 Myr history of the Earth it has been formed at a rate of about 1.7 cubic kilometers per year – in 1 Myr that's 1.7 million cubic kilometers, equivalent to a cube 120 km on its side. It is eroded by various processes, and some of the eroded material becomes ocean sediments. A small proportion of these is subducted (Figure 2.13), but the sediments are too buoyant for much to be removed in this way. It is not clear whether other processes could be significant. Overall, it is thought that the volume of continental crust plus sediments derived from it, has increased over much of Earth history.

The underlying and ultimate cause of plate motion is convection in the mantle, driven by the heat left over from the formation of the Earth 4,600 Myr ago, and the heat released by the decay of long-lived radioactive isotopes. But how can a *solid* mantle convect? It might surprise you to learn that at the temperatures and pressures in the mantle convection *does* occur, though on a timescale of many thousands of years.

This has been a thumbnail sketch of some of the processes of plate tectonics (Greek "tekton" = "builder"), a global system unique in the Solar System, and responsible for most of the sculpting of the Earth's surface. Though brief, it will suffice for our considerations of life on Earth and on other planets.

Atmosphere and oceans

Silicates and iron would never, on their own, have produced life on Earth. Other substances are essential, even if present only as traces at and near the Earth's surface. The oceans and the atmosphere are major repositories of such traces. The oceans consist almost entirely of water, but with gases dissolved from the atmosphere.

Atmospheric compositions are usually expressed as percentages. In the Earth's atmosphere today, the nitrogen molecule (two atoms of nitrogen, N-N, written N_2), comprises 77% of all the molecules present, the oxygen molecule O_2 comprises 21%, water vapor (H_2O) 1% on average, argon 0.9%, plus numerous traces present in very small proportions. These traces include carbon dioxide

FIGURE 2.14 The Earth from space, which shows that the atmosphere is a thin veneer. (NASA/JPL-Caltech, part of PIA01321)

(CO_2), a greenhouse gas (see below), and ozone (O_3), that shields the Earth's surface from ultraviolet radiation from the Sun, which is damaging to many life forms, including humans. Note that in the distant past the composition of the atmosphere was quite different, as you will shortly see.

On each square meter of surface at sea level there stands a column of atmosphere with a mass of 10,300 kg, but because gases are so compressible the density declines rapidly with altitude, from about 1.23 kg per cubic meter at sea level to about 0.09 kg per cubic meter at just 20 km altitude. For comparison, liquid water at the Earth's surface has a density of about 1,000 kg per cubic meter. The atmosphere is thus a thin, insubstantial veneer. This is apparent in Figure 2.14. Space is just 100 km or so above our heads, a distance along the Earth's surface that can readily be covered in an hour.

Among the atmospheric gases dissolved in the oceans are O_2 and CO_2. Without dissolved O_2 many aquatic life forms would die, just as many surface dwellers, including us, would die without atmospheric O_2. The oceans are intimately coupled to the atmosphere, physically and chemically, and account for most of the water above the base of the Earth's crust (there might be traces in the mantle). Liquid water is far less compressible than gases, and so the density of the oceans does not increase much with depth from its surface value.

The greenhouse effect

When I mentioned atmospheric CO_2 I expect that your mind turned to global warming. Before the Industrial Revolution the atmospheric fraction of CO_2 was a bit less than 0.028%. It is now 0.037% and rising fast. It is widely thought that this is due mainly to human activities, notably forest clearance and the burning of fossil fuels. Global warming is linked to the rise in CO_2 through the greenhouse effect.

A greenhouse effect arises whenever there is one or more atmospheric constituent that absorbs little of the incoming solar radiation but more of the infrared radiation emitted by the surface. It is as if the planet has been wrapped in a blanket – the surface temperature is greater than it would otherwise have been.

The Earth's surface temperature, averaged over the year and over the surface is 15°C. If there were no greenhouse effect this would be –18°C, and over much of the Earth's surface it would be too cold for life. The greenhouse effect is not a bad thing! About 21°C of this 33°C difference is due to water vapor, most of the remainder being due to CO_2. Note that it is two minor atmospheric constituents that create the blanket. It is the increase in the CO_2 content that is thought to be causing most of the rise in temperature across much of the globe. The predicted rises are not huge – a few °C over the next 100 years – but they *will* affect our lives.

In the distant past, when the Earth was young, the greenhouse effect might explain why the Earth seems not to have been frozen at that time, despite the lower luminosity of the young Sun at that time (Section 2.2). At that distant time the Earth's atmosphere might have been richer in greenhouse gases than it is today. This is one solution to what is called the faint Sun paradox.

2.4   THUMBNAILS OF THE OTHER BODIES IN THE SOLAR SYSTEM

In the broadest terms, the other terrestrial planets, and the Moon, have a bulk composition that resembles the Earth, with an iron rich core, partially molten (though in some cases probably to a far smaller extent than the Earth's core), and a solid silicate mantle. Mercury has a large iron core in proportion to its size, and is thus greatly enriched in iron compared to the Earth. The Moon is greatly depleted in iron. Neither Mercury nor the Moon has an appreciable atmosphere.

Venus resembles the Earth in size and fairly closely in bulk composition, but it has a far more massive atmosphere, and one with a quite different composition, being dominated by CO_2. Under this massive blanket the greenhouse effect maintains a global mean surface temperature of an astonishing 470°C. Mars has a thin atmosphere, also dominated by CO_2. The resulting weak greenhouse effect sustains a global mean surface temperature of only –55°C. It has a bulk composition only modestly depleted in iron compared to the Earth. Both Mars and Venus are devoid of oceans and lakes – Mars is too cold, Venus is too hot.

The asteroids have a range of compositions, from a high enrichment in iron, through an iron silicate mix, to a high enrichment in silicates with materials rich in compounds of carbon. Mars's tiny satellites, Phobos and Deimos, just a few kilometers across, might be captured asteroids.

The giant planets, particularly the two largest, Jupiter and Saturn, have a composition much closer to the Sun than to the terrestrial planets. They are fluid throughout and are both dominated by hydrogen and helium, roughly in solar proportions. They differ from the Sun in having a higher proportion of the other elements, 5–10% in Jupiter and perhaps 2–3 times more in Saturn. These other elements are mainly in the form of icy materials, notably water (H_2O) and (liquid) rocky materials. Note that the term "icy materials" refers to a range of substances that are icy at sufficiently low temperatures i.e., they resemble water ice in appearance. Though water is the most abundant icy material in the Solar System, methane (CH_4), ammonia (NH_3), CO_2, carbon monoxide (CO) and N_2 are also present in significant quantities, particularly beyond the asteroids.

In the case of Saturn, the icy and rocky materials seem to be mainly concentrated into a core, but in Jupiter the concentration seems to be less, perhaps far less. This might be due to Jupiter's higher internal temperatures, that increase with depth to reach central values modeled at around 20000°C, whereas in Saturn the models give central temperatures around 9000°C.

Uranus and Neptune differ from Jupiter and Saturn in having less hydrogen and helium. These elements constitute massive atmospheres that overlie liquid cores, which, as in Saturn (and Jupiter if it has a core) consist largely of water and rocky materials.

The many satellites of the giant planets have compositions ranging from rocky bodies, dominated by rocky materials and iron, to icy-rocky bodies, in which icy materials (mainly water) are at least as prevalent as rocky materials and iron, and in many cases are dominant. Materials rich in carbon are present in most rocky and icy-rocky satellites. In all but the largest icy-rocky bodies the icy materials are solid. The Edgeworth–Kuiper belt objects (including Pluto) and the Oort cloud objects are rich in icy materials, and also have a significant proportion of carbon rich materials. Comets are derived from these populations and thus have similar compositions.

Planets, satellites, and the other bodies in planetary systems beyond ours, are expected to display a broadly similar range of compositions.

But how did the Solar System come into being? The better we can answer this question, the better we will be able to understand the other systems.

2.5   THE ORIGIN OF THE SOLAR SYSTEM

The Sun and the planets formed about 4,600 Myr ago from a cloud of gas and dust. The gas consisted mainly of hydrogen and helium, and the dust mainly of the other 90 chemical elements. As the cloud contracted, its rotation quickened and it flattened to form a thick disc (Figure 2.15). The dust settled through the

FIGURE 2.15 An artist's impression of the disc of gas and dust from which the Sun and the planets formed. The Sun is forming at the center. This is an oblique view – the disc is circular. (Julian Baum, Take 27 Ltd.)

gas to the mid-plane of this disc, to form a thin dust disc. In its inner region, near where the Sun was forming at the center of the disc, the dust disc consisted mainly of iron, silicates and other compounds with high melting points. Further out, where temperatures were lower, carbon-rich compounds were also present in solid form. Even further out, beyond what is called the ice line, as its name suggests, water ice (and other icy materials) was present in the dust.

The Earth and the other terrestrial planets formed from the inner region of the dust disc, which is why iron and silicates dominate their composition. The first step in planet formation was the agglomeration of some of the dust to form kilometer-sized bodies called planetesimals. These collided, causing disruption but also growth. The outcome was a set of what are called planetary embryos, ranging in diameter from a few hundred to a few thousand kilometers. Over perhaps 100 Myr, by far the slowest stage in the process, the set of embryos interacted until just a few large bodies were left – the terrestrial planets. The energy liberated in their formation caused much of each planet to melt, to form an iron core surrounded by a silicate mantle.

There were also small quantities of other substances. In the case of Venus, Earth, and Mars these subsequently gave the planets their early atmospheres, and at least in the case of the Earth, its oceans. A proportion of these substances came with the dust and larger bodies throughout the formation of the terrestrial bodies, and a proportion came near the end of their formation, from bodies rich in water and carbon compounds, analogous to present day comets and certain asteroids. These late arrivals contributed to a heavy bombardment, which in the case of the Earth continued to about 3,900 Myr ago i.e., to about 700 Myr after the Earth had acquired nearly all of its mass.

In the outer Solar System things were different, because beyond the ice line water ice was present. This, along with rocky materials, allowed kernels to form with several times the mass of the Earth, massive enough to capture hydrogen and helium gas from the disc. This gas contracted to form massive hot envelopes surrounding the hot cores. This is the favored model for the formation of the giant planets.

Capture ceased when the disc around the Sun was dissipated by a burst of intense solar activity called the T Tauri phase. Uranus and Neptune formed more slowly than Jupiter and Saturn because of their greater distance from the Sun and the consequent lower density of the disc. However, at their present distances from the Sun their cores would only have formed after the disc was dissipated, which would have left them bereft of the thick atmospheres rich in hydrogen and helium that we know they possess. Fortunately there are computer simulations that show that Uranus and Neptune could have formed closer to Saturn than they are today, and thus would have been able to capture gas, though less than Jupiter and Saturn, as observed. They then migrated outwards, through a variety of subtle gravitational interactions with the remnants of the disc and with other bodies. Planetary migration is an important feature of many other planetary systems, as you will see in Chapter 11.

A planet failed to form in the space between Mars and Jupiter because of the gravitational disturbance caused by Jupiter that halted further build up. The asteroids thus formed instead. The present population of asteroids is different from the original population, which has been modified by collisions between asteroids. This has changed the size distribution. Mass has also been lost, not only through collisions but also through the gravitational influence of Jupiter, and to a lesser extent Mars.

Planetary satellites are formed in a variety of ways. The Earth's Moon is probably the result of the impact of an embryo, similar in size to Mars, late in the Earth's formation. Some of the mass of this body would have joined the Earth, the rest would have formed a disc of debris around the Earth, from some of which the Moon coalesced. Mars's two tiny satellites are probably captured asteroids. The inner satellites of the giant planets, which includes all the big ones, are thought to have formed from a disc of dust and rubble around the planet, somewhat like the formation of the terrestrial planets around the Sun. Many of the other satellites, particularly the small ones far from their planet, are thought to have been captured from interplanetary space.

All the four giant planets are encircled by rings, that of Saturn being presently the most extensive by far (Figure 2.4). The rings consist of small solid particles, in a thin sheet fairly close to the planet. The particles are tiny fragments of larger bodies that came sufficiently close to the giant planet to be disrupted by its powerful gravity. Particles are gradually lost from the rings, but a large supply of new particles could arise from the disruption of a small satellite that strays too close to the giant. Presumably this happened relatively recently in Saturn's vicinity, which is why it presently has the most extensive ring system.

After the formation of the planets, there was plenty of material left over, from dust, to small rubble, to larger sizes. In the inner Solar System the composition would have been predominantly rocky. In the outer Solar System icy materials would also be present. Pluto, and many of the other Edgeworth–Kuiper belt objects, must be bodies that were formed beyond Neptune but never grew further, presumably because of a shortage of material. Other Edgeworth–Kuiper belt objects, and the Oort cloud objects, are thought to have been flung outwards in the late stages of giant planet formation by the giants' gravity.

That concludes my outline of the Solar System and how it formed. This has set the stage for the discussion of life on Earth, to which I now turn. This, in turn, will set the stage for the search for life beyond the Solar System.

3

The nature of life on Earth

We are, of course, very familiar with life on Earth – its variety, its distribution, its behavior. This is the *only* sample we have of life in the Universe, and so we look to it for guidance as to what it is we should look for beyond the Earth, and where we are most likely to find it. This cannot be achieved in a few words, which is why I've devoted three chapters to life on Earth – its nature in this chapter, its origin in the next, and its evolution in the one after. Throughout Chapters 3–5 I will concentrate on what is needed for discussing extraterrestrial life in later chapters.

3.1 THE CELL

Life on Earth comprises the biosphere, the assemblage of all things living and their remains. Figure 3.1 shows a typical scene of life on Earth. It is neither herds on an African plain, nor a rainforest, nor an ocean teeming with fish, but single celled creatures – unicellular creatures – that are only 1–100 micrometers across, and thus need to be viewed under a microscope if we are to see much structure (recall that a micrometer is a millionth of a meter). The cell is the basic unit of all life on Earth, an enclosed environment within which the processes of life are conducted. The membrane that encloses the cell regulates the exchange of substances between the cell and its environment.

There is a great variety of unicellular creatures, with bacteria as a very large group. There are also multicellular creatures. If these consist of just a few cells, such that a microscope is needed to see them, then, along with unicellular creatures, they are called microbes. Larger creatures are the ones familiar to us, such as insects, plants, and animals. The human body contains about ten million million cells, aggregated into various organs – heart, liver, brain, and so on. Supposing a typical cell to be 10 micrometers across, this number of cells would stretch for 100,000 km if laid in a row.

Figure 3.2 shows the essential components of the two types of cell that constitute all life on Earth – the prokaryotic cell and the eukaryotic cell (Greek: "karuon", "kernel"; "pro", "before"; "eu", "good"). In unicellular organisms the cell can be of either type, whereas almost all muticellular organisms consist of eukaryotic cells. The prokaryotic cell is the simpler of the two. It consists of an

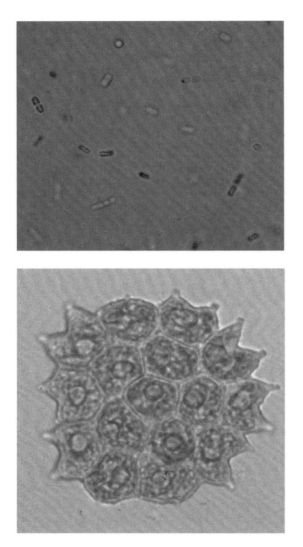

FIGURE 3.1 A typical scene of life on Earth. Upper image: the bacterium *Synechococcus* sp., where the single cells (each about 4 × 2 micrometers) have not always separated after dividing. Lower image: a group of non-bacterial single cells of an organism called *Pediastrum boryanum* (each cell about 10 × 8 micrometers). (Derek Martin)

enclosing membrane made of proteins and a particular type of lipid (the technical terms in Figure 3.2 and in these paragraphs are explained in Section 3.2). Outside the membrane the great majority of prokaryotes have a cell wall that provides a measure of rigidity. The membrane encloses a fluid called cytosol that consists mainly of salty water containing proteins. Within the cytosol floats the genetic material DNA and small particles called ribosomes that contain RNA.

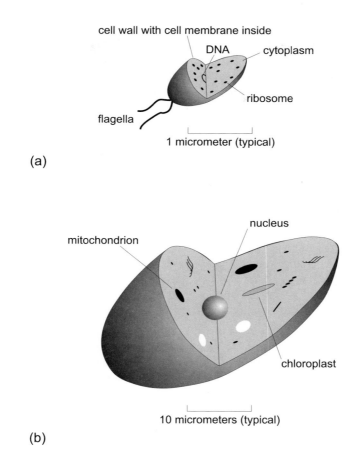

cell wall with cell membrane inside

DNA

cytoplasm

ribosome

flagella

1 micrometer (typical)

(a)

nucleus

mitochondrion

chloroplast

10 micrometers (typical)

(b)

FIGURE 3.2 The essential components of (a) the prokaryotic cell (b) the eukaryotic cell.

This mixture inside the cell is called the cytoplasm. Most prokaryotic cells can move by means of protein strands attached outside the cell wall – these are called flagella.

The eukaryotic cell is typically much larger that the prokaryotic cell, 10–100 micrometers instead of about 1 micrometer. It is also much more complex. Of particular note are the various structures in the cell called organelles. One such is the nucleus, where much (but not all) of the cell's DNA is housed. Another is the mitochondria (singular, mitochondrion) inside which cell respiration ("breathing") occurs. In green plants there are chloroplasts, which are the sites of photosynthesis. Mitochondria and chloroplasts also contain DNA. There are various other organelles and other structures that will not concern us.

In multicellular creatures different cells take on different functions. For example, in animals a nerve cell carries out a different function from a muscle cell or a liver cell. Most plant cells have a rigid wall made of cellulose fibres held together by a glue. By contrast cells in mammals have no such wall and generally change shape readily.

Table 3.1 shows the composition of a typical mammalian cell in terms of the broad types of chemical compounds that it contains. Though the cells of other creatures have somewhat different proportions, the Table gives the correct impression that any living cell consists mainly of water, with proteins as the next largest component. Then there are various components present in much smaller quantities. The next section is devoted to these various chemical components.

TABLE 3.1   Chemical components of a typical mammalian cell

	Water	Proteins	Lipids	Poly-saccharides	RNA and DNA	Others
Mass (% of total)	70	18	5	2	1	4

3.2   THE CHEMICAL COMPONENTS OF LIFE

In terms of the chemical elements, life is dominated by hydrogen (H), oxygen (O), carbon (C), and nitrogen (N), with representative percentages (by numbers of atoms) 63% H, 28% O, 7% C and 2% N. At significant fractions of a percent are calcium (Ca) and phosphorus (P), plus traces of many other elements. These are not present as a random jumble – far from it, as you will see. First, consider water.

Water

This is made up of a relatively simple molecule that consists of one atom of oxygen (O) bound to two atoms of hydrogen (H), written as H_2O. Figure 3.3 is a simplified diagram of a water molecule, where H and O denote the positions of the centers of the hydrogen and oxygen atoms – the tiny atomic nuclei. The space between these nuclei is filled with the orbiting electrons that surround the nuclei. Water, as exemplified by Table 3.1, accounts for most of the mass of a living cell.

Water is essential for all life as we know it. In its liquid form it dissolves a great range of substances enabling them to be transported inside the cell, and also, in multicellular creatures, between cells. It is also a participant in biochemical reactions, when it splits into OH^- and O^+. The "–" sign indicates that the OH molecule (called hydroxyl) has one electron too many to be electrically neutral – it is an example of a negative ion. The "+" sign indicates that the O atom has one electron too few to be electrically neutral – it is an example of a positive ion. All cells require water, and require it to be liquid inside them over at least part of the cell's life cycle.

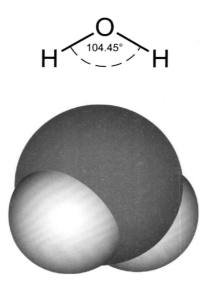

FIGURE 3.3 The structure of a molecule of water – simplified. The ball and stick version shows the structure more clearly than the ball version, though the latter is a better representation of how atoms fill space. The O–H distance is 95.54 millionths of a micrometer.

Proteins

Proteins are compounds of carbon and a few other elements. Almost all of the biochemicals of life are compounds of carbon. These belong to a vast family called organic compounds, carbon being the essential component. Nearly all are large molecules i.e., they contain many atoms. Among all the chemical elements, carbon is the best at forming huge, complex molecules that are the basis of life.

About 100,000 different proteins have been identified in terrestrial organisms, and between them they fulfill a wide range of functions – structural, transport, storage and catalytic. A catalyst is a substance that greatly increases the rate of a chemical reaction, but which emerges unscathed after it has done its job.

A protein is made up of units called amino acids. Figure 3.4 shows the general molecular plan of an amino acid (note that the bonds between the atoms are not, as shown here, in the same plane). They differ in the nature of "R", which can range from a single hydrogen atom (to give glycine) to a few tens of atoms, with C and H predominant. A protein consists of about 50–1,000 amino acids of various types, joined together in a string. This is an example of a polymer – a large molecule made of many much smaller molecules strung together. In many polymers the small molecules are identical; in others, as in proteins, they have a "family resemblance".

In some proteins several strings are intertwined but they remain string-like; these are called fibrous proteins, and their function is structural. In other

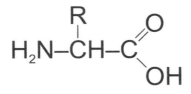

FIGURE 3.4 Amino acids, the building blocks of proteins. Amino acids differ in the nature of R. Note that, unlike the simplified layout shown here, the atoms are *not* all in the same plane.

proteins a single string is wound to become roughly spherical. Many of these act as catalysts of biochemical reactions i.e., they speed up the reactions. Such catalysts are called enzymes, and without them biochemical reactions would occur far too slowly to sustain life, or would not outpace competing non-biological reaction rates sufficiently for life to survive. Nearly all enzymes are proteins. The precise geometrical form of an enzyme determines which biochemical reactions it catalyses.

About 20 different amino acids make up all the proteins found in life on Earth. With an "alphabet" of about 20 letters and a "word" that is 50–1,000 letters long, you can probably see that the 100,000 or so different proteins that have been identified in terrestrial organisms are an extremely small fraction of those that could exist. This is to be remembered when we consider possible alien biologies.

RNA and DNA

These are large, complex organic compounds that are at the heart of protein synthesis, and are central to the processes by which organisms reproduce themselves.

RNA – ribonucleic acid – is a long molecule, another polymer. A typical segment is shown in Figure 3.5. The spine consists of a string of identical units. Each unit contains an organic molecule called ribose – a member of the sugar family, a family of compounds of C, H, and O. This is attached to a phosphate – compounds of P, O, and a metal. Attached to this spine are what are called bases. There are four bases, labelled A, C, G, and U, standing for adenine, cytosine, guanine, and uracyl (and not for chemical elements). These are fairly small, with numbers of atoms ranging from 13 to 17. All but adenine are compounds of C, N, H, and O – adenine lacks O. A single segment of RNA, called a nucleotide, consists of a spinal unit attached to a base. The nucleotides can be joined in any order – the sequence of the segment in Figure 3.5 is written -U-A-C-A-G-. A molecule of RNA consists of a string of anything from hundreds to tens of thousands of nucleotides, and so a huge variety of RNA sequences is possible. The string is usually folded.

DNA – deoxyribonucleic acid – is a relation of RNA, with three main

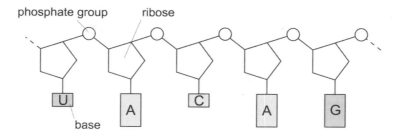

FIGURE 3.5 A segment of an RNA molecule.

differences. First, in the spine, ribose is replaced by a different sugar, deoxyribose. Second, the base uracyl does not occur, but another base thymine (T) is present instead. Third, there are two strands joined by the bases to form a "ladder" with "rungs". This is illustrated in Figure 3.6(a), which shows a typical segment of a DNA molecule. Each rung is either A-T (or T-A) or C-G (or G-C). A-G, for example, does not occur. The rungs can occur in any order, and so, with thousands of rungs in a particular DNA molecule, there is a huge number of possible base-pair sequences. The ladder is twisted to form the famous double helix, as illustrated in Figure 3.6(b). The double helix is not straight, but curled up, to form a complex three dimensional structure.

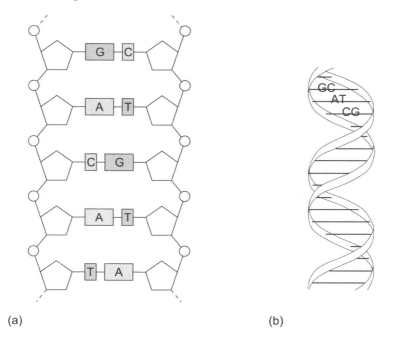

(a)   (b)

FIGURE 3.6 (a) A typical segment of a DNA molecule. (b) The famous DNA double helix.

The remaining components

The remaining components in Table 3.1 are the lipids (fats and oils), the polysaccharides (many sugar molecules strung together – another polymer), and "others", comprising small organic molecules, and inorganic ions (e.g., derived from common salt when it is dissolved in water). A few specific compounds will make brief appearances below.

You can see that an apt name for life on Earth is carbon-liquid water life. This name highlights that large carbon compounds are the basis for life, and that liquid water is essential.

How do these components work together – how does life live, and how does it reproduce?

3.3   THE FUNDAMENTAL PROCESSES OF LIFE

Any living organism on Earth is involved in three processes. First, biosynthesis, in which small organic molecules are constructed from even smaller molecules and atoms, and then combined to form the much larger molecules found in the cell – notably, proteins, RNA, DNA, lipids, and polysaccharides. Second, reproduction, in which cells make copies of themselves, so that life is sustained from one generation to the next. Third, catabolism in which large molecules are broken down into smaller ones.

In each of these processes energy is transferred from one place to another. In this respect, life is like every other process in the Universe – nothing can happen without energy transfer, the energy often changing from one form to another. Important forms of energy for life are chemical energy, thermal energy, the energy in radiation (such as solar radiation), and the energy in motion (kinetic energy). If there are no energy sources, there can be no life.

Creating the cell's energy stores

There is a large variety of ways in which organisms store energy. Energy is stored in chemical form in sugars, polysaccharides, certain lipids, and, rarely, proteins. Where do these stores come from? For some unicellular organisms, and many multicellular organisms the answer is "food". This is organic material ingested or absorbed from other organisms. Humans are particularly omnivorous, obtaining organic matter from a great variety of sources. Organisms that rely on such ready-made organic material are called heterotrophs ("feeders on others"). Other organisms build energy stores from simple inorganic compounds – these are called autotrophs ("self-feeders"), and they lie at the base of the food chain.

Green plants are autotrophs that manufacture organic compounds through the process of photosynthesis, in which solar radiation at red and blue wavelengths initiates a complex sequence of reactions that convert CO_2 and water into a sugar called glucose. Land plants get the CO_2 from the atmosphere, and the water from the soil (rarely from the atmosphere). Plants that grow in water get CO_2 that has been dissolved in the water around them. The chemical energy in the sugar has come from the energy in solar radiation. Other substances are produced subsequently, such as amino acids (that build proteins). A by-product of this form of photosynthesis is the oxygen molecule O_2, released into the atmosphere (or oceans), so this is called *oxygenic* photosynthesis. It generates nearly all the O_2 in the Earth's atmosphere.

Green plants are multicellular organisms. Some unicellular organisms also photosynthesize in the above manner. Some do it differently, notably a few prokaryotes that perform photosynthesis by using other molecules in place of water. In this case there is no O_2 by-product. It is less efficient than oxygenic photosynthesis in that it captures a smaller proportion of the available solar radiation, though it can be utilized by organisms that are intolerant of oxygen.

Some autotrophic prokaryotes create energy stores using chemical reactions that do not involve photosynthesis. This is called chemosynthesis. In chemosynthesis the organism takes advantage of substances in its environment that can be made to react within the cell and release energy that is used to create energy stores. Even in clear water, at depths greater than only about 100 meters, there is insufficient sunlight for photosynthesis and therefore only chemosynthesis is possible. The same is true in underground caves and crevices, and in crustal rocks. One type of organism releases methane during chemosynthesis – these methanogens will feature in our discussion of the detection of extraterrestrial life.

Utilizing the cell's energy stores

Energy is obtained from the stores through catabolism (the breaking of large molecules into smaller ones). In almost every organism, unicellular or multicellular, catabolism happens through a process called respiration, which in animals is also called breathing. Respiration can be aerobic or anaerobic.

Aerobic respiration relies on O_2, the second most abundant component of the Earth's atmosphere, and also present in the Earth's rivers, lakes, and oceans. O_2 is used in a series of chemical reactions that convert the organic materials in the energy store to water and carbon dioxide, releasing energy in the process. Most eukaryotes are aerobic.

In anaerobic respiration, as you might guess, oxygen is not used. This is essential for organisms intolerant of oxygen. It is carried out by a variety of unicellular organisms, both prokaryotes and eukaryotes. A familiar example is the fermentation that produces alcohol through the activity of yeasts, which are colonies of unicellular eukaryotes. In this case the respiration results in the production of CO_2 as before, but instead of water, alcohol is produced (more accurately called ethanol). Anaerobic respiration yields much less energy from a store of given size than aerobic respiration.

Organisms that use aerobic respiration are called aerobes. The rest are anaerobes, and they divide into those for which oxygen is toxic and those that will use aerobic respiration when oxygen is available.

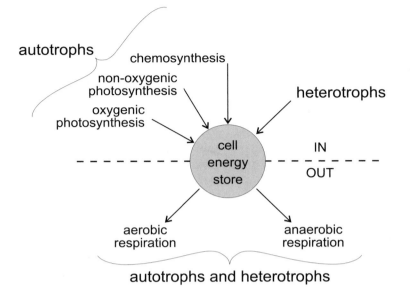

FIGURE 3.7 Types of organism, and their means of energy storage and release.

Figure 3.7 summarizes and links the various processes and types of organism that have been introduced in this section.

Let's now turn to more biosynthesis in the cell – protein synthesis, leading on to reproduction and mutation.

Protein synthesis

After water, and excluding cell walls in some organisms, proteins make the largest contribution to the mass of the cell (Table 3.1). They also have the greatest number of functions as I outlined earlier. The one thing that a protein molecule cannot do (with very few exceptions) is to make a copy of itself. Therefore, to form a protein something else is almost always needed. That "something else" is the DNA in the cell. The DNA contains the instructions for making all the cell proteins, and as such is the "blueprint" of the cell – it is where all the genetic information resides. Protein synthesis, in most cells, requires more energy than any other biosynthetic process.

Stripped to its bare essentials, protein synthesis is accomplished when a segment of the DNA blueprint is read with the aid of enzymes and various forms of RNA, including that in ribosomes. The reading of a specific segment results in the creation of a specific protein. In heterotrophs most of the amino acids that make up a protein have to come from food that traverses the cell membrane, though animals can synthesize some of them. In autotrophs the amino acids are synthesized within the cell.

Reproduction and mutation

All organisms have limited life-spans. It is therefore essential for the survival of a particular type of organism that it reproduces itself. This applies as much to bacteria as it does to oak trees, elephants, and us. In all organisms today the information for making all of its many and varied proteins is contained entirely within its DNA. This is tantamount to saying that the DNA houses the information for making a copy of the organism. It thus follows that the persistence of a type of organism from one generation to the next requires the passing on of its DNA.

In prokaryotes reproduction involves the splitting of a cell into two cells, each cell having the same DNA. Thus, one molecule of DNA has become two. How does this happen? The two strands in the DNA are separated by enzymes. This leaves two sets of exposed, unpaired bases. Each exposed base, with the aid of an enzyme, captures the nucleotide with the complementary base from free nucleotides that are present in abundance in the cell's cytoplasm – these free nucleotides will have been manufactured through enzyme action. In this way two identical DNA molecules are created, each the same as the original molecule.

The copying of the DNA is not always perfect. The "children" are therefore not always identical to the parent. If the "children" are viable this will lead to an increase in variety. Such an increase can also result from a prokaryotic cell transferring some of its DNA to another, possibly of a different type, such that the recipient's DNA is modified. Another source of variation is DNA damage. This can be caused by chemical attack, by solar ultraviolet radiation, and by energetic particles and gamma rays from space. Yet another cause is processes internal to the cell that cause segments of DNA to be moved to another location in the molecule. All these DNA changes are called mutations.

In eukaryotic cells, cell division also occurs. One process, called mitosis, enables unicellular eukaryotes to reproduce by division, and enables multicellular organisms to grow, and to replace dysfunctional or lost cells. However, the process of producing a *new* organism can be different. In plants and animals this is usually by sexual reproduction, in which half the DNA comes from special cells in the female, and the other half from special cells in the male. In animals these are eggs and sperm respectively. This means that the DNA in the offspring differs from that in each parent, though it is sufficiently similar that we recognize the offspring as belonging to the same species as the parent – oak trees produce oak trees, elephants produce elephants, humans produce humans. The cell division that results in this mixing of DNA from two parents in eukaryotes is called meiosis.

Sexual reproduction clearly promotes variation in the DNA in the offspring. Further variation results from mutations by all the processes outlined for prokaryotic cells.

Thus, in all organisms DNA is not an invariable entity passed from one generation to the next. It is this that has enabled life to evolve from its earliest forms to the huge variety of organisms that we see today. Evolution is discussed in the next chapter.

The chicken and the egg

You have seen that that the whole process of DNA copying is driven by enzymes. But enzymes are proteins, and DNA is required to manufacture proteins. This is reminiscent of the famous chicken and egg conundrum – which came first, the chicken or the egg? This question is of particular importance to the origin of life on Earth, as you will see, also in the next chapter.

3.4   LIFE IN EXTREME ENVIRONMENTS

We exist at the Earth's surface, and without clothing and fire we would be confined to places where liquid water is available in drinkable quantities, where there is a food supply, and where temperatures are rarely outside the approximate range 5–45°C (41–113°F). With clothing and fire, and dwellings, we can live almost anywhere on the solid surface of the Earth. The very highest altitudes are excluded because the atmosphere is so thin that there is insufficient oxygen for sustained respiration. Other places are excluded, for example, because of noxious volcanic gases.

Humans thus exist in non-extreme environments. Well, we would wouldn't we? It is we who have labelled our habitats as non-extreme. To some other creatures, as you will see, it is we who inhabit extreme environments.

The notion of a non-extreme environment extends beyond what humans can inhabit. Sustained temperatures can be as high as about 60°C without being regarded as extreme. Also, although we cannot live in water, we do not regard depths down to a few hundred meters as extreme. Nor do we regard anaerobic environments as extreme – these are readily found close to home e.g., in marshland.

Table 3.2 lists some extreme environmental conditions under which terrestrial organisms are known to live. Organisms that can live in such conditions are called extremophiles. Note that they must be able to *live* in the environment i.e., grow and reproduce, and not just survive until conditions ameliorate. Indeed, extremophiles live best under extreme conditions. The categories in the table are not mutually exclusive – the same organism can live under more than one extreme. Most extremophiles are prokaryotes.

High temperatures

The extremes in Table 3.2 are striking. Consider high temperature. Organisms that live at temperatures above about 80°C are called hyperthermophiles ("lovers of extreme heat"). To live at 80°C is impressive enough. Yet the current high temperature record is 121°C, held by a unicellular prokaryote with the unflattering provisional name, Strain-121.

At sea level pure water boils at about 100°C (212°F). So how can cells have liquid water above this temperature? You are probably familiar with the fact that

Table 3.2 Some extreme environmental conditions under which terrestrial organisms live.

	Limit(s)	Type of organism
Temperature	−18 to 15°C	psychrophiles
	60 to 80°C	thermophiles
	80 to 121°C	hyperthermophiles
Pressure[1]	0.0061 bar to a few bar	(no special name)
	up to 1,300 bar[2]	piezophiles (barophiles)
Salinity	15–37.5% NaCl	halophiles
pH	0.7–4	acidophiles
	8–12.5	alkalophiles

(1) A bar is the average atmospheric pressure at sea level on Earth.
(2) In the oceans.

water boils at *less* than 100°C at *low* pressures. On the top of Everest, 8,848 meters (29,028 feet) above sea level, the pressure is about 30% that at sea level, and water boils at about 70°C. Conversely, at high pressures the boiling temperature is raised. In the oceans, even at 1 km depth, the pressure is high enough to raise the boiling temperature to about 300°C. This is the case at even shallower depths on land, because of the greater density of the crust.

Hyperthermophiles thus live in high pressure environments – they are also piezophiles. To reach the temperatures at which hyperthermophiles live some source of heat is needed, and at many locations across the Earth this is provided by volcanic activity. It is at such hot locations that hyperthermophiles are found.

Strain-121 was discovered on the ocean floor at a depth of about 3 km, about 400 km off the coast of Washington State, USA, at a hydrothermal vent (a "black smoker") called Finn. At such vents volcanically heated water pours out, much of it recycled ocean water, enriched in dissolved gases. Figure 3.8 shows a hydrothermal vent, the water darkened by the formation of solids in it as it meets the cool ocean water.

The upper temperature limit for carbon-liquid water life could be as high as about 160°C. Above this temperature essential carbon compounds are broken down. Sufficiently large pressures are found at depths of only about 50 meters in the oceans, and rather less in the Earth's crust, because of its greater density.

Low temperatures

At the other temperature extreme, psychrophiles ("lovers of cold") are known to be able to live down to temperatures of about −18°C (Table 3.2). To do so they must avoid three threats. Most obviously, the water in the cell must not freeze. Pressure does not help – it hardly affects the *freezing* temperature. But 0°C is for *pure* water. Certain proteins lower the freezing point of water in the cells of psychrophiles. Second, the rates of biochemical reactions must not become too low. These rates decline rapidly as temperature falls, and below about −10°C they

FIGURE 3.8 A hydrothermal vent (a "black smoker") about 1 meter tall, pouring out water at temperatures up to about 400°C, and containing dissolved gases, including CO_2, H_2, CH_4, and H_2S. (Dudley Foster, Woods Hole Oceanographic Institution)

are extremely low. Psychrophiles avoid this problem via special enzymes, not found in other cells, that speed reactions at low temperatures. Finally, the cell membrane must not become too rigid to function as a regulator of what substances pass through it. This is avoided by the incorporation of special lipids into the membrane that keep it flexible.

Pressure

In Table 3.2 the pressure is given in bars. Atmospheric pressure at sea level rarely deviates by more than 2–3% from 1 bar (1,000 millibars). The lower end of the pressure range in Table 3.2, 0.0061 bar, is determined by the requirement to have water as a liquid. Below 0.0061 bar, pure water is stable only as a solid or as a gas. For comparison, the pressure at the top of Everest is about 0.3 bar. Organisms could live at pressures below 0.0061 bar if the cell wall could maintain a sufficiently high pressure at the cell temperature. However, the lowest ambient pressure at which life has been found is about 0.01 bar, where a prokaryote called Bacillus Subtilis has been discovered.

The current upper end of the pressure range of organisms known to live at high pressure (piezophiles) is around 1,300 bar in the oceans. Such pressures are found in the deepest part of the Mariana Trench, in the Western Pacific Ocean. This is the deepest the ocean gets anywhere on Earth – 11.0 km below sea level. Multicellular microbes as well as unicellular microbes have been found at such depths.

For unicellular creatures the true upper limit to pressure could be far higher – higher pressures occur just a few kilometers down in the crust. Water-filled pores and cracks at such depths are plenty big enough for single cells to live. Unicellular creatures – mostly prokaryotes – have been found at depths of a few kilometers in the crust. The water contains chemical reactants and a variety of autotrophs are found, relying on chemical energy. The pores and cracks are not isolated, so exchange of water and cells must occur. Some estimates of the *mass* of the organisms in the deep crust exceed estimates of the total mass of organisms at shallow crustal depths, plus those in the oceans and at the surface. The *number* of tiny organisms well beneath our feet would then be truly vast. They would not be packed together – a depth range of several kilometers provides plenty of volume.

Unicellular organisms do not die when brought from depth up to the surface, nor do those living at the surface when taken down. By contrast, multicellular organisms perish, particularly when the pressure change is rapid.

Salinity, acidity, alkalinity, ionizing radiation

Table 3.2 also lists three other types of extreme environment. Salinity is a measure of the amount of salt in the environment, mainly common table salt (sodium chloride, NaCl), but other salts too. We live in a virtually salt free environment e.g., fresh water has a salinity of less than 0.05% by mass. Even sea water has an average salt content of only about 3%. Halophiles ("salt lovers") are a small group of unicellular creatures that live in salinities from 15–37.5%. All cells need salts (Section 3.2), but the salinity of the cytoplasm in some halophiles is only the few percent that is typical of other cells. These cells have various ways of expelling salts from the cell. Other halophiles survive by adapting the cell machinery to high salt content – these cells perish outside a salty environment.

Acidity and alkalinity are measured in units of pH. Pure water, which is neither acid nor alkali, has a pH of exactly 7. Acids have smaller values, the more corrosive the lower the value. The pH of vinegar (a solution in water of acetic acid) is between 2.0 and 3.5. Table 3.2 shows that acidophiles live in environments that span the pH of vinegar, and even down to a pH of 0.7. Battery acid, which is sulphuric acid in water, has a pH of 0.5, so corrosive that it would quickly destroy our skin! Alkalis have pH values greater than 7. Alkalophiles live in alkaline environments, such as soda lakes and chalky soils. Acidophiles and alkalophiles live in their extreme environments through processes in their cells that move the pH in the cell towards 7.

One of several extremes not listed in Table 3.2 is heavy exposure to ionizing radiation. This comprises fast atomic particles, and ultraviolet, X-ray, and gamma radiation. Strong ionizing radiation at the Earth's surface is found near radioactive materials, naturally occurring and concentrated by human activity, such as in nuclear power stations. Some prokaryotes can live in levels of radiation that would quickly be lethal to us. Most notable is a bacterium called *Deinococcus radiodurans*, which withstands radiation levels 100–1,000 times greater than that which would kill us. It is thought that this ability is a by-product of its adaptation to extremely dry conditions (another type of extreme environment).

The existence of extremophiles on Earth shows us that wherever there is a remote chance of carbon-liquid water life occurring, it does occur. It is a remarkable fact that wherever carbon-liquid water life *could* exist on Earth, it *does* exist. This surely increases the likelihood that if we find habitable extraterrestrial locations, they will in fact be inhabited.

3.5 ALIEN BIOLOGIES?

All life on Earth is carbon-liquid water life. Could life elsewhere have a different basis? Could water be replaced by another liquid? Could the complex compounds required to contain the genetic information have a basis other than carbon?

In the absence of any observational evidence there has been much speculation about the biochemistry and appearance of alien life forms. With regard to appearance, much of the speculation has had no roots in science (Figure 3.9). But in recent decades this has changed, and as a result, Chapter 15, the final chapter in this book, is devoted to what might the aliens be like – speculative, but rooted in science.

I think that he is running a temperature.

FIGURE 3.9 The aliens have landed! A case of misidentification. (Graham Read)

4

The origin of life on Earth

How did the enormous variety of life on Earth originate? This is one of the great unsolved mysteries in science. It might come as a surprise to you that we simply don't know. In the next chapter, you will see that we do have a fair understanding of how life has evolved *since* its origin. But as to the origin itself, we lack sufficient evidence to uncover how it happened. We are left with little more than a variety of speculations. This is regrettable, because it hampers our ability to establish the likelihood of life arising elsewhere.

But life clearly did originate, and as there seems to be nothing special about the Earth as a planet, astrobiologists, including me, think that there could well be life on other Earth-like planets.

What do we know about the origin of life on Earth?

4.1 THE LAST COMMON ANCESTOR

When we look at the history of life on Earth, we find that the oldest organisms are single celled prokaryotes. These are also the simplest organisms. You might be surprised to learn that there is ample evidence that all life on Earth today sprang from a single unicellular prokaryote! This is called the last common ancestor.

Why is there just one? Can it really be that all life on Earth has evolved from one organism? Why can't we have many ancestors? This is ruled out by the observational evidence. All life on Earth today is biochemically very similar – all proteins are made from the same 20 or so amino acids, and all RNA and DNA have the same sets of nucleotides. The same is true of fossils, as seen in the rare cases when their biochemicals are sufficiently intact. There are many other similarities across the biosphere, at the biochemical and cellular level. The descendants of another common ancestor would surely be distinct in some way. But at present there is no evidence for such life, nor any evidence in the fossil record. If they did exist, even beyond the time of the last common ancestor, we have found no surviving traces.

Therefore, as far as we can tell, there *is* a unique common ancestor, and it must have been biochemically very similar to all life on Earth today. It is likely to have been a unicellular prokaryote. This is the simplest form of life, and simplicity surely precedes complexity. This idea is supported by the fossil record, which shows that prokaryotes extend further back in history than eukaryotes.

The origin of life thus comes down to the emergence of our last common ancestor from the endowment of chemicals carried by the Earth at its birth. But I'll first address the less thorny question of how long ago it happened.

How long ago?

Figure 4.1 shows a timeline of Earth history, focusing on life. As described in Chapter 2, the Earth was born around 4,600 Myr ago, from embryos, planetesimals, and dust in a disc around the Sun. At first the surface was hot, and subject to a violent bombardment as the last planetesimals were swept up, as illustrated in Figure 4.2. This heavy bombardment lasted until about 3,900 Myr ago, when it was declining rapidly. It is possible that, in the 700 Myr between 4,600 Myr and 3,900 Myr, there were intervals of light bombardment during which life originated, but it could not have survived the later heavier bombardments. These dates, and others in Figure 4.1 have been obtained by a technique called radiometric dating.

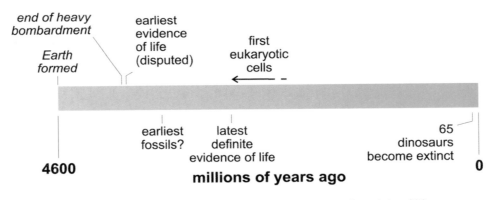

FIGURE 4.1 The timeline of Earth history, focusing on the origin of life.

Box 4.1   Radiometric dating

Radiometric dating depends on isotopes that are unstable in that their nuclei emit atomic particles or gamma rays that results in the transformation into other nuclei, often of a different chemical element. Such isotopes are said to be radioactive. Examples of atomic particles emitted by the nucleus are neutrons and electrons. The electrons result from the conversion of a neutron into a proton. (A neutron carries no electric charge; it is electrically neutral, whereas a proton has one unit of positive electric charge and an electron one negative unit. Therefore, overall electric charge is neither created nor destroyed in this conversion.)

Consider the example of the isotope of the chemical element rubidium, Rb. This element has 37 protons in its nucleus. It has just one stable (non-radioactive) isotope with a nucleus containing 48 neutrons, making 85 nuclear particles in total. It is thus called rubidium-85, written in shorthand as ^{85}Rb. Rubidium has four radioactive isotopes, the most abundant of these being ^{87}Rb i.e., 50 neutrons in the nucleus. This nucleus emits an electron. As a result the number of neutrons in its nucleus has gone down by one and the number of protons has gone up by one. The total number of particles in the nucleus is still 87, but now comprised of 49 neutrons and 38 protons. By definition it is no longer a rubidium nucleus. It is, in fact, a nucleus of strontium, ^{87}Sr. This isotope is stable, so this is the end point of what is called the radioactive decay of ^{87}Rb.

The emission of the electron from ^{87}Rb, as with all nuclear emissions, is random in time. A particular nucleus of ^{87}Rb could emit an atom straight away, or in a year, or in a time longer than the 4,600 Myr age of the Solar System. The best we can do is to specify the time it takes for half of a large number of radioactive nuclei to decay into another nucleus. This is called the half life of the nucleus. In the case of ^{87}Rb it is 48,000 Myr.

It is tempting to conclude that, with half the ^{87}Rb nulei gone in 48,800 Myr, the whole lot will be gone in a further 48,800 Myr. This is not the case. The 50% left after 48,800 Myr is the starting point for the next 48,800 Myr. At the end of this time a further 50% are gone, leaving 25% of the original number. The number of ^{87}Rb nuclei versus time is thus a smoothly descending curve, approaching zero only after very many half lives. This curve is, in fact, an example of an exponential curve.

How is this used to obtain ages? We must first be certain of the nature of the event we are trying to date. One event is when the mineral containing rubidium became chemically closed i.e., no longer exchanged isotopes with its environment. The simplest case is when, at closure, there was no ^{87}Sr in the mineral. Thus, by measuring the present day ratio of ^{87}Sr to ^{87}Rb, and by knowing the half life of ^{87}Rb, we can determine what fraction of a half life has elapsed since closure, and hence obtain the age of the mineral. (In the case of ^{87}Rb the half life is considerably longer than the age of the Universe, so a good deal less than a full half life will have elapsed.)

In practise there will probably have been ^{87}Sr in the mineral at closure. This complication is overcome by examining minerals with different initial endowments. The details will not concern us.

Many other radioactive decays are used to date events in Earth history, often involving several stages before the end point in a stable isotope. One such is the decay of the isotope of uranium, ^{238}U, to the stable isotope of lead, ^{206}Pb (from the Latin "plumbum" for lead). The overall half life is 4,470 Myr.

The first evidence we have of life being present on Earth dates from about 3,850 Myr ago, in metamorphosed (pressure and heat modified) sedimentary rocks at Isua, western Greenland. This is a remarkably short time after the decline of the heavy bombardment. Though there are no fossils in these ancient rocks there are chemical signatures from carbon isotopes. In particular the ratio of the

FIGURE 4.2 An artist's impression of the heavy bombardment of the Earth, which persisted from the origin of the Earth 4,600 Myr ago, until about 3,900 Myr ago. (Julian Baum, Take 27 Ltd.)

amount of the carbon 12 isotope to the carbon 13 isotope is slightly higher than the ratio in carbon known to come from non-biological sources. Modern cells that photosynthesize produce such an enhancement, though there are also *non-biological* ways of producing it, so the biological interpretation is somewhat controversial.

Among the oldest fossils are stromatolites. Figure 4.3(a) shows a present day stromatolite. It consists of layers of minerals laid down by colonies of bacteria called cyanobacteria (these are among the bacteria that generate oxygen by photosynthesis). These flourish in shallow, warm water. Figure 4.3(b) shows a fossil stromatolite. The oldest stromatolite fossils date from 3,460 Myr ago, from Warrawoona in Australia. As well as a superficial resemblance to present day stromatolites, they also have structures that resemble fossilized bacteria. Even though non-biological formation of the Warrawoona stromatolites is possible, most scientists believe them to be fossils. Similarly, some structures in South Africa's Baberton greenstone belt, 3,300–3,500 Myr old, are thought by most

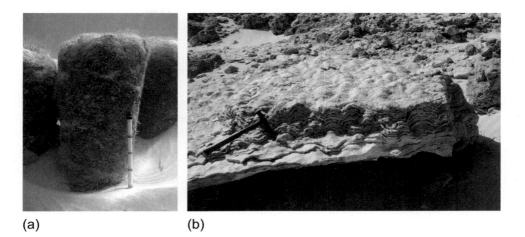

(a) (b)

FIGURE 4.3 (a) A present day stromatolite in shallow seas around Lee Stocking Island in the Bahamas. The divisions on the rod are 100 mm long. (R P Reid, University of Miami) (b) A fossil stromatolite about 15 Myr old, at Wadi Kharaza, Egypt. (Brian R Rosen)

scientists to be fossils of prokaryotic cells. The earliest undisputed evidence for life is fossil cyanobacteria from Pilbara Australia, at 2,700 Myr, and stromatolites of the same age from Fortescue, also in Australia.

Clearly, the earliest common ancestor must predate 2,700 Myr, and probably predates 3,460 Myr, or possibly even 3,850 Myr. Infall from space in the form of volatile rich planetesimals and bodies resembling comets must have delivered significant quantities of molecules that presumably aided the emergence of carbon-liquid water life on Earth. Evidence for this infall comes in the form of meteorites. There is a range of types, and in one of these (called carbonaceous chondrites) we see a wide range of organic compounds, including amino acids. Observations of comets indicate that they too contain organic compounds.

Certainly, by the end of the heavy bombardment around 3,900 Myr ago, the Earth had its complement of the building blocks of biomolecules, oceans, and an atmosphere of some sort, and was ready to produce life.

With the last common ancestor probably predating 3,460 Myr ago, it is reasonable to conclude that the Earth is likely to have given birth to its last common ancestor within about 1,000 Myr of the Earth's formation. Such timing might well apply to other planets that resemble the Earth.

But how do we get from simple molecules to a cell?

4.2 THE ORIGIN OF THE LAST COMMON ANCESTOR

Granted that the last common ancestor was a single prokaryotic cell, possibly more primitive than those we see today, it would have carried its genetic information in DNA. It would also have relied on proteins to fulfill the wide variety of functions performed by proteins today. But such a cell could not have sprung directly from simple building blocks (such as amino acids) – this is far too great a step.

A major difficulty is the "chicken and egg" problem, briefly mentioned in Chapter 3. Proteins require the genetic information in DNA for their synthesis from small molecules, yet proteins as enzymes are needed to promote this synthesis. Furthermore, proteins as enzymes are needed to copy DNA during reproduction. Thus, we can't have DNA without proteins and we can't have proteins without DNA.

RNA world

One way out of this dilemma was suggested in the late 1960s, independently by several scientists. It was proposed that the precursor of life's last common ancestor used RNA rather than DNA as the store of genetic information. Crucially, the RNA in this theory replicates without the aid of proteins, and catalyzes all the chemical reactions necessary for the precursor to survive and reproduce, including the synthesis of proteins. The world in which organisms rely on RNA in this way is known as RNA world. The basis for this proposal was as follows.

- The nucleotides in RNA are more readily synthesized than those in DNA.
- It is easy to see how DNA could evolve from RNA and then, being more stable, take over as the repository of genetic information.
- It was difficult to see how proteins could replicate in the absence of nucleic acids (though it has recently been discovered that some proteins can self-replicate).

The proposal was further strengthened in 1983 when it was discovered that RNA could act as a catalyst. Unfortunately this is only to a very limited extent, and there is consequently a huge gulf between what we know RNA can do, and what it would be required to do in RNA world.

Assuming that RNA world existed, it would have been a large step towards the origin of life. But it begs the question of how we get the components of RNA – the bases, the sugar ribose, and the phosphates. It is difficult to make these in any plausible early atmosphere, particularly some of the bases and ribose. Infall from space does not help – it brings no ribose, no bases.

Worse, even if we had all the components of RNA present, there is the huge problem of assembling them into nucleotides (a single unit of RNA), and assembling the nucleotides into an RNA strand with the order of 1,000 bases required as a basis for life. An additional difficulty would have been the likely abundance of molecules related to the RNA nucleotides. Their incorporation into RNA would have destroyed its functionality. RNA world thus faces many problems.

The formation of cells

As well as the biochemicals of life, we also need to provide a protected environment – a cell. Quite how the first true cell appeared, like much else, is unknown, but droplets called coacervates might provide a clue. When a solution of lipids is shaken, droplets of higher concentration are formed, bounded by membranes. Small molecules such as amino acids can diffuse through the membrane and join to form short strings that cannot escape. This is a mechanism for providing a protected environment for producing long biological polymers. However, it is a long way from a living cell. Like much else in origin of life studies, coacervates give us a plausible hint, but no more.

The role of minerals

The difficulties faced by RNA world have led to the hypothesis that the surfaces of minerals, and tiny compartments in minerals, played an essential role in the early development of life. Tiny compartments can shelter simple molecules, and surfaces can promote the combination of these molecules, ultimately to form long biological polymers. It is also the case that a surface can acquire a concentration of molecules that is much higher than that in the surrounding environment, thus facilitating the build-up of larger molecules from smaller ones.

Clays are central to the mineral hypothesis because of their intricate surfaces and labrynthine interior structures (Figure 4.4). Clays form through the recrystallization of silicates and other minerals dissolved in water, and consist of minerals modified by the inclusion of water molecules and the hydroxyl (OH) fragment. There is plenty of evidence that the Earth had a large amount of liquid water at its surface 3,800 Ma ago, for example from the metamorphic rocks at Isua in Greenland, and there are observational and theoretical reasons to suppose that liquid water was present far earlier. It seems certain that there was liquid water before any RNA world, and that therefore clays were also present at that time.

There is experimental evidence for the potential importance of clays to the origin of life. When a water-based solution of amino acids is evaporated from a vessel containing clays, the surfaces of the clays build short chains that resemble proteins of the sort found in cells today. The similarity with evaporation from a shallow pond or tidal pool with a muddy bottom on the early Earth must be noted. Clay surfaces have also been found to promote the assembly of RNA from its constituents. In these examples some selectivity of the clays is apparent – many types of complex molecules *could* have been synthesized, but biomolecules are prominent. This selectivity is characteristic of other minerals with a complex architecture. It is also possible that the enclosure of biomolecules within a cell was first accomplished on or within a mineral. Alternatively, biomolecules could have escaped from minerals and were then trapped in coacervates that had formed in the surrounding watery environment.

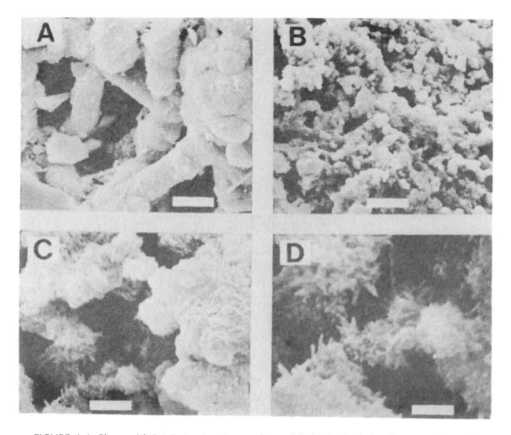

FIGURE 4.4 Clays with intricate structures of possible biological significance. The scale bars are as follows: (a) 2 mm, (b) 10 mm, (c) 2 mm, (d) 1 mm. (M Raymond Bayan)

In recent years attention has turned to the role of minerals deep under the Earth's surface. On the ocean floors, at hydrothermal vents (Section 3.4), simple minerals such as the iron oxide called magnetite (Fe_3O_4) and the iron sulphide pyrite (FeS_2) might be able to accomplish much. Magnetite could have catalyzed the formation of ammonia (NH_3) from the H_2 and N_2 that these vents emit, thus providing the NH_3 that biochemical reactions involving nitrogen require. More profoundly, groups of minerals, particularly iron sulphides and nickel sulphides, could have acted as templates, catalysts, and as the energy sources that produced the first biomolecules, and enabled them to form quasi-living systems. The minerals themselves might have contributed some of their atoms to various biomolecules. The high temperatures near such vents are not a problem, because biomolecules can exist inside minerals at temperatures greater than they can outside.

In the 1970s, Graham Cairns-Smith of the University of Glasgow, UK took the role of clays to the limit, by proposing that clays, not carbon compounds, provided the first genetic material. It was *life*, though not as we know it. However, the consensus now is that though clays (and other minerals) might well have played a role in the origin of life, there was never any mineral-based genetic system.

So there we have it. Not a satisfactory point to leave our discussion of the origin of life, but take comfort from the fact that it appears to have happened fairly soon after the end of the heavy bombardment. If life can appear with such facility then our hopes are raised that extraterrestrial life exists throughout the cosmos.

4.3 CHIRALITY IN BIOMOLECULES

One feature of life on Earth that would aid identification of extraterrestrial life is associated with chirality. Figure 4.5 shows the amino acid alanine – note that the various components of the molecule do not lie in the same plane. Two forms are shown, labelled L and D, which differ only in that they are mirror reflections of each other. This means that the molecule cannot be superimposed on its mirror image, regardless of how you twist and turn them. Likewise, if you have the backs of your hands facing you, and then move one hand to cover the other, they do not match. In the case of molecules, such a molecule is said to be chiral.

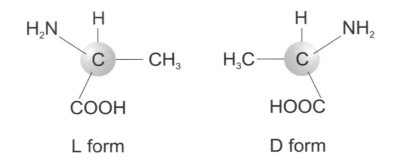

FIGURE 4.5 The amino acid alanine in L and D forms. The sphere around the central carbon atom is to help indicate the three dimensional bond directions. L and D stand for "left" and "right", derived from the Latin for left ("laevus" – "leave-us") and right ("dexter").

A striking feature of life on Earth is that in all DNA and all RNA, the sugars in the spines occur *only* in the D form, and very nearly all amino acids in proteins occur *only* in the L form. The other forms are of no biological use at all. For example, a ribosome constructed to assemble L amino acids will not attach D forms – the external shape of the D form is wrong, rather as a left-handed glove will not fit a right hand. Why should some biomolecules have to be D, others L? Minerals offer one possible solution.

Some minerals are also chiral, in that they have faces that are mirror reflections of each other, for example, the faces of a crystal of calcite ($CaCO_3$). It has been found that if calcite is exposed to amino acids, one type of face acquires an excess of the L form on its surface, and the other type acquires the D form, particularly if the surfaces are finely terraced. The two types of face are equally abundant, so we need to assume that it was by chance that the L form appeared first and through replication quickly dominated the biosphere. Another example is provided by certain clays, where, due to geometrical space limitations, a surface can acquire the property of either being able to adsorb only the L form of a chiral molecule, or only the D form. Again it would seem to be up to chance which chiral form came to dominate our biosphere.

There are other explanations of chiral biomolecules that do not involve minerals. One is that circularly polarized light bathed the cloud from which the Solar System formed. In this case a slight excess of one hand of molecule over the other could have been produced in the prebiotic molecules. Circularly polarized light can be crudely modeled as waves spiraling through space like a corkscrew, twisting either clockwise or anticlockwise, and the favored molecular hand depends on which sense the light happened to have during the molecule's formation in space. The light in star-forming regions has been observed to be circularly polarized, and some of the amino acids in meteorites rich in carbon compounds have a slight excess of one hand over the other. However, this explanation, like that involving minerals, is no more than plausible. The origin of chirality in biomolecules is yet another mystery.

4.4   WHERE DID LIFE ORIGINATE?

It is perhaps natural to think that life must have originated in the oceans, near the surface within reach of sunlight and perhaps in shallow pools or tidal flats where small component molecules could have been concentrated by evaporation, perhaps on clay surfaces. There are however four reasons for us to look deeper.

First, the heavy bombardment must have frustrated the origin of life at the surface many times. If life originated beneath the surface it would have been protected from this and other environmental hazards, and we get several hundred Myr of opportunity denied to life on the surface.

Second, photosynthesis was not operating at the origin of life – it is a very complex process that must be some distance along in evolution. It is possible that a primitive system existed in which solar radiation was used as a source of energy, but it is more likely that a non-solar source powered the earliest life. There are biologically potent energy sources well away from the surface. These are based on chemical reactions between materials coming from deep in the Earth, and materials in the environment into which they emerge, such as at hydrothermal vents (Section 3.4). Therefore, although most life on Earth now relies ultimately on photosynthesis, this was surely not the case at the beginning.

We thus need not confine ourselves to the surface of the Earth in looking for life's origin.

Third, much of the biosphere today, perhaps most of it by mass, is well below the surface of the solid Earth, in the crust (Section 3.4). We need not assume that this deep biosphere spread from the surface – it could have been the other way.

Fourth, among the earliest forms of life still with us today, is a large group of prokaryotes that are thermophiles, flourishing in the sort of temperatures found at hydrothermal vents and deep in crustal rocks. Though these organisms must have evolved from the last common ancestor, and though colonization by species that originated in cooler environments cannot be ruled out, they do indicate that life might have originated at high temperatures.

For all these reasons the view that life originated deep in the Earth and spread to the surface has gained considerable support among biologists, regardless of the extent of mineral involvement. An origin near the surface cannot be ruled out, but it currently seems less likely.

By removing the origin of life from where the Sun keeps a planetary surface suitably warm, the number of potential habitats in the Solar System increases. This is also the case for other planetary systems, as you will see in later chapters.

One other possible source of life on Earth is extraterrestrial. In modern times this idea dates back to the early twentieth century, to the Swedish physical chemist Svante August Arrhenius, and the British physicist Lord Kelvin. They each proposed the idea of panspermia in which dormant cells travel through interstellar space, seeding planets. In Lord Kelvin's version the dormant cells are carried inside meteoroids, and are thus protected from radiation. Panspermia doesn't solve the problem of how life arose – it merely transfers the origin elsewhere – though it does give more time and space for life to originate.

It seems to be a viable idea, at least within the Solar System. The possibility that life could have emerged on our planetary neighbors, and that we have nearly 40 meteorites that seem to have come from Mars, raises the intriguing possibility that we are all Martians!

However it happened, as soon as a self-replicating carbon-based biomolecular system emerged, it began to use up the environmental resources on which it relied. Evolution would then have led to greater efficiency and greater complexity, until the last common ancestor arrived. Subsequently, there would have been a negligible chance of any other self-replicating, evolving system emerging – life as we know it would have been a powerful "predator", occupying all possible niches.

Regardless of when and where the last common ancestor appeared, how has life evolved since then? This is the subject of Chapter 5.

5

The evolution of life on Earth

5.1 THE TREE OF LIFE

Figure 5.1 is the tree of life, which shows the relationship between the various life forms on Earth. As we go down the tree we are reaching further back in time. This is a highly simplified version, but one that is perfectly adequate for our purposes. The full version, from which Figure 5.1 has been derived, has been established by comparing the biological dissimilarity (distance) between different species of organisms by comparing their RNA and DNA. For example, domestic cats are a species close to the various wild species of cat (e.g., leopards) in that they have almost the same DNA and RNA. By contrast, there is a greater difference between domestic cats and dogs, and an even greater distance between domestic cats and any species of oak tree. Distances can be measured between all organisms, unicellular and multicellular.

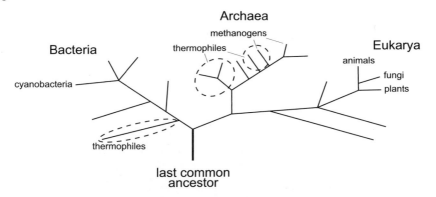

FIGURE 5.1 The tree of life. As we go down the tree we are reaching further back in time.

Though there might have been life, now extinct, before the last common ancestor, it is this ancestor that stands at the base of the tree. From this ancestor have sprung all subsequent life forms. These are divided into three domains, the Bacteria, the Archaea ("ar-keeya"), and the Eukarya ("you-carrya"). The Eukarya comprise all organisms that consist of one or more eukaryotic cells. The Bacteria and Archaea are organisms that consist of prokaryotic cells, nearly all of them unicellular.

Until the late 1970s there were only two domains – the Bacteria and the Eukarya. Then the molecular biologist Carl Woese showed that in certain bacteria a certain type of RNA is different from the RNA that performs the same functions in cells of other bacteria. In 1977 he proposed that the Bacteria should be divided into two domains, the Bacteria and the Archaea. By the mid-1980s, as evidence grew, his proposal had become widely accepted.

Box 5.1 Carl Richard Woese

Carl Woese is an American molecular microbiologist, who, in 1977, on the basis of RNA evidence, split the domain Bacteria into two domains, the Bacteria and the Archaea. This view gained acceptance only slowly, and there was hostility from some biologists and even from some non-scientists. Even today there is a small minority that does not accept this split. This is characteristic of revolutions in science – as evidence piles up showing the accepted model or theory to be inadequate, a fairly sudden switch occurs, with the great majority of scientists switching to the new model/theory. Those still adhering to the earlier model/theory, gradually die out.

Carl Woese (born 1928) in 2004. (Don Hamerman)

Almost all Bacteria and Archaea are microscopic organisms, and include the great majority of extremophiles. It is among the Eukarya that, as well as microscopic organisms, we find nearly all of the large organisms, by which I mean anything that can be seen easily without the aid of a microscope. Among the smallest of these are mites, for example the house dust mite measuring about 0.4 mm long and about 0.3 mm wide – you need good eyes and favorable lighting to see one. The mites are animals, one of the four so-called kingdoms of the Eukarya, called Animalia (Figure 5.1). The other kingdoms are Plantae (plants), Fungi (e.g., yeast and mushrooms), and Protoctista, an "all other eukaryotes" kingdom comprising a wide variety of forms. Most Protoctista are unicellular, though seaweeds are a familiar multicellular example.

Within each of the three domains are many species that have become extinct. For example, among the animals in the Eukarya (Figure 5.1) we have the whole range of dinosaurs, which became extinct about 65 Myr ago, and the dodo, a single species of bird that lived on Mauritius and became extinct as recently as the late seventeenth century. Extinctions are part of the process of evolution, as you will see shortly. But all species in the three domains, extant and extinct, have their origin in the last common ancestor.

Ascending the tree of life

Time advances as we move up each branch. The first branching above the last common ancestor shows the Bacteria dividing from the Archaea. It is not known which of these two domains of life extend towards the last common ancestor.

Some of the branching dates in Figure 5.1 are known. These have been obtained from fossils embedded in or between radiometrically dated rocks. For example, plants appeared by about 470 Myr ago. The first animal fossils date back to about 575 Myr ago. Other branching dates are not known, or have large uncertainties, including the ancient branching into Bacteria and Archaea. The later branching of Eukarya could have occurred as early as 2,700 Myr ago, or as late as 2,100 Myr. There are many more branchings than the handful shown.

Each of the branches bifurcates time after time until we end at individual types of organism. In the case of plants and animals these are called species. Two plants or two animals belong to different species if a fully developed male of one species and a fully developed female of the other cannot produce a fully fertile hybrid under natural conditions, or if its production is extremely rare. With living creatures it is often impossible to put this to the test, and impossible with fossils. Comparisons of appearance, and, if possible, of behavior, can help to distinguish species, as can comparisons of RNA and DNA where available.

There is a hierarchy of classification. Species are grouped into a genus, followed by successive groupings into order, class, phylum ("file-um"), and kingdom. For example, the domestic cat (all varieties) is classified as follows: species *Catus* (domestic cat); genus *Felis* (wild and domestic cats); family Felidae (all cats); order Carnivora (carnivores); class Mammalia (mammals); phylum Chordata; kingdom, Animalia. The phylum specifies a characteristic body plan.

For example, the Chordata have backbones, and so include fish, frogs, and humans, but not spiders. There are 24 phyla in the kingdom Animalia (Figure 5.1 uses the less formal name, "animals").

In the genus *Homo* (Latin for "man") there is only one species alive today – us. We are all *Homo sapiens* (Latin for "wise man" – very self-congratulatory!). *Homo sapiens* emerged from earlier species in Africa about 190,000 years ago, perhaps earlier. This is only 0.19 Myr into the past, far, far less than the time ago that life first emerged on Earth. We are very recent arrivals.

Up to about 24,000 years ago (at least in Europe) there was certainly one other species of *Homo, Homo neanderthalensis* (from the Neander Valley in Germany), commonly known as Neanderthal Man. Full blown neanderthal characteristics appeared about 130,000 years ago, in Europe and parts of western Asia. It is thought that they became extinct in the face of competition from *Homo sapiens* as we spread into the colder climates that they inhabited. Some anthropologists believe that there was a measure of interbreeding, in which case we have a small genetic inheritance from the neanderthals.

It must be emphasized that, as well as the appearance of new species, other species have become extinct. This results in branches that do not reach to the present. Extinctions also lie within the branches that *do* reach us today. These are earlier species that have led to modern forms.

We come now to another big question. By what mechanism did the tree of life develop from the last common ancestor? In other words, what is the process of evolution?

5.2   THE PROCESS OF EVOLUTION

In Section 3.3 it was stated that DNA is not an invariable entity passed from one generation to the next. In the great majority of cases this either has no discernible effect, or a tiny effect of no importance, or a damaging effect that results in infertility of the offspring, death of the offspring, or maladaption to the environment that leads to extinction of the line after a few generations. But some descendants will be more suited to the environment than others, and it is these that will have a better chance of survival and reproduction. The inherited characteristics that promote survival to reproduce will therefore result in growing numbers, limited only by the availability of resources such as food. Gradually, new species emerge.

By this means, life has evolved from its earliest forms. Evolution has enabled the survival of certain species, the emergence of new, viable species, and the extinction of others. Indeed, it has been estimated that over the whole of Earth history 99% of species have become extinct, never to appear again. Evolution has led to the huge range of species on Earth, unicellular and multicellular, each adapted to its own environment.

It is important to note that evolution usually takes place slowly over many generations. For example, many people wonder how the eye could have evolved.

Box 5.2   Charles Robert Darwin (1809–1882)

Darwin was an English naturalist, who developed the theory of evolution by natural selection – "survival of the fittest". Much of Darwin's field evidence that led him to this theory was acquired by Darwin during the second voyage of HMS Beagle, 1831–1836. This took Darwin westwards around the world. During the five year voyage he spent over three years ashore making detailed observations of plants and animals, and collecting fossils.

Charles Darwin at 51, soon after publication of his epoch making book in 1859.

In 1844 he sketched his conclusions for his own use. This sketch contained the seeds of his theory, but he only went public in 1858, prompted by a memoir sent from the Malay Archipelago by the Welsh naturalist Alfred Russel Wallace (1823–1913). In this memoir essentially the same theory as Darwin's was outlined. As a result, on 1 July 1858, a paper by Darwin and a paper by Wallace were read at a meeting of the Linnaean Society in London, though neither man was present.

It took Darwin over a year to condense his vast mass of notes into the epoch making book, "The Origin of Species by Means of Natural Selection", which was published in November 1859. It is truly a book that has revolutionized our view of the living world.

This did not happen in one step, but in many small steps over a long time. The eye is thought to have started with a light sensitive patch that gave a survival advantage. Perhaps a hollow developed over subsequent generations – the beginnings of an eye socket that protected the light sensitive patch. Later, a lens evolved. So important is vision to survival that it has evolved independently in widely different phyla of Animalia, giving us a great variety of forms, such as the eyes of mammals, flies, and spiders.

Charles Darwin

Long before it was known that DNA was at the root of the variation in offspring, Charles Darwin, on the basis of observations of a huge number of species, proposed, with Alfred Wallace, the theory of evolution by natural selection. In this theory variations in organisms occur, and those organisms best suited to their environment will have the greater number of offspring and thus will flourish. The word "theory" should not detract from the mass of observational evidence that supports it. Darwin's theory has evolution as a slow, gradual process.

Since Darwin's time it has been discovered that, as well as slow evolution, there have been short intervals of more rapid evolution, usually consequent upon rapid, mass extinctions, as you will see in Section 5.3.

5.3   MAJOR EVENTS SINCE THE LAST COMMON ANCESTOR

Figure 5.2 shows the timeline of Earth history, focusing on life since the last common ancestor. This ancestor must predate the earliest evidence we have of life, be this as early as 3,850 Myr ago or possibly as late as 2,700 Myr. Some unknown time after this ancestor was alive the split between the Archaea and Bacteria occurred. At those distant times the Earth was very different from today. One major difference was in the composition of the atmosphere. The Earth acquired its atmosphere (and oceans) from some combination of outgassing of the rocks that made up its bulk composition, and bodies rich in water, other icy materials, and carbon compounds, that constituted the heavy bombardment of the Earth in its first 700 Myr or so. Most models of this early atmosphere have it dominated by CO_2 and N_2, plus significant amounts of water vapor and smaller quantities of CO and CH_4. It is certainly the case that O_2 was a mere trace.

You saw in Section 4.1 that the only evidence we have that life existed at 3,850 Myr is carbon isotope ratios, and that this evidence is controversial – non-biological explanations are possible. If the isotope ratios are indeed from life, then this would indicate that photosynthesis was already in operation. However, it is unlikely that it was generating oxygen (oxygenic) – there are photo-synthesizing microbes today that don't generate oxygen (non-oxygenic). Also, there is biological evidence that oxygenic photosynthesis evolved through a combination of two non-oxygenic microbes. Better evidence for oxygenic

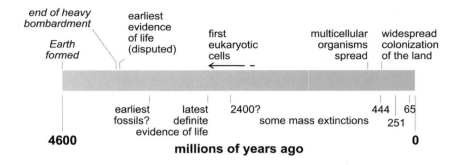

FIGURE 5.2 The timeline of Earth history, focusing on life since the last common ancestor, and showing some mass extinctions.

photosynthesis is at around 3,500 Myr ago, based on the stromatolite-like fossils of that date, provided that they are the products of cyanobateria (as are modern stromatolites). It is by 2,700 Myr ago that oxygenic photosythesis was well established, as indicated by cyanobacteria fossils and stromatolite fossils of that date (Section 4.1).

Arise, Eukarya!

At about this time, an event of central importance in the evolution of life might have been taking place – eukaryotic cells might have been emerging. The range of times indicated in Figure 5.2 indicate the likelihood that Eukarya emerged no earlier than 2,700 Myr ago, and possibly as late as 2,100 Myr ago, with a preference for earlier times. The complex structure of the eukaryotic cell seems to have arisen through symbiosis, in which two or more prokaryotic cells united. There is evidence for this among the organelles in eukaryotic cells (Section 3.1). For example, the chloroplasts that photosynthesize in plant cells, and the mitochondria that carry out aerobic respiration in plant and animal cells, have DNA that show bacteria to be their closest relatives. The prokaryotes that united included anaerobic forms, but also aerobic forms, and consequently many eukaryotic cells were able to utilize oxygen for aerobic respiration. Today, almost without exception, eukaryotic cells require oxygen, either in the air or dissolved in water, whereas some prokaryotes are killed by oxygen, others can tolerate it, and only a small proportion require it.

Nearly all multicellular forms of life are made from eukaryotic cells, and their proliferation was the next great event in the timeline of life on Earth. It occurred as late as about 610 Myr ago, when the previously very rare multicellular organisms became much less rare, larger and more complex (Figure 5.2). Corals and creatures resembling jellyfish appeared, but also quilted creatures that have no modern counterparts. "Shortly" afterwards, at 545 Myr, there was a huge increase in the number and variety of such organisms, and also in their range of habitats.

FIGURE 5.3 A fossil trilobite, *Ogygiocarella*, from just after the Cambrian period in central Wales. It is 75 mm long. (Peter Sheldon)

This rather precise date defines the beginning of what is called the Cambrian period (545–495 Myr ago). Many of these new organisms had hard parts, such as shells and exterior skeletons, and they have therefore been readily preserved as fossils. Figure 5.3 shows a fossil of one of the many species of trilobite, which flourished in the oceans from the early Cambrian. Though there were environmental changes occurring at the time, it is not clear that such changes were the causes of the Cambrian "explosion" in the diversity of eukarya. Further contributory factors might have been the increase in atmospheric oxygen content (see below), and the generation of new and diverse habitats through the migration of continents across the Earth's surface, resulting from plate tectonics.

Up to the Cambrian, nearly all life had been confined to the oceans. At earlier times there had been isolated bacterial mats on land, even as far back as 2,700 Myr, though they were sparse. But it was not long into the Cambrian before life with its enlarged potential for diversity was able to colonize the land far more widely. This began around 490 Myr, just after the end of the Cambrian period, but the main colonization was between 440 Myr and 420 Myr.

By 420 Myr ago all the phyla of multicellular organisms were on land and in the sea. Our own genus, *Homo*, emerged just 1.5–2.5 Myr ago, and our own species, *Homo sapiens*, the only one remaining species in the genus *Homo*, as you saw earlier, emerged a mere 0.19 Myr ago (190,000 years)! Evolution continues. Many modern species of shark date back about 100 Myr. Can humans look forward as a species to such longevity? We have no idea.

Mass extinctions

As well as the *gradual* extinction of species, there have also been several episodes of what are called mass extinctions, when a large percentage of the existing species became extinct in a relatively short time. Each mass extinction provided "ecological space" that new species could inhabit, and therefore each one is characterized not only by a huge loss of species, but in huge numbers of new species. This has led to the term punctuated evolution.

The two most dramatic mass extinctions occurred 251 Myr ago and 65 Myr ago (Figure 5.2). The first one defines the end of the Palaeozoic ("old life") Era, when about 90% of species became extinct. This is called the Great Dying. The one at 65 Myr defines the end of the Mesozoic ("middle life") Era, when 70% of marine species became extinct, plus a large proportion of land species, including, famously, the dinosaurs. In each case it took 0.1–1 Myr for all the species to disappear – a mass extinction is not an overnight phenomenon. It took roughly ten times longer for the biosphere to recover.

In the past 500 Myr there have been three rather modest mass extinctions, at 444 Myr, 360 Myr, and 210 Myr, plus a few small ones. There were probably mass extinctions well before 500 Myr, but the evidence for these is less secure, partly because of the less complete fossil record.

The biosphere took tens of millions of years to recover from each mass extinction. A few branches of the tree of life were terminated. Other branches were much depleted, such as the trilobites, which were almost wiped out in the Great Dying – their only living descendants are woodlice! New branches appeared, such as the dinosaurs, which arose about 30 Myr after the Great Dying.

There was a large element of chance in which species survived a mass extinction, though species that had a broad diet or could survive the new climatic conditions, had a significant advantage.

Overall, as a result of evolution, punctuated or otherwise, the diversity and complexity of life on Earth has generally increased throughout much of Earth history.

The causes of the mass extinctions at 65 Myr and 251 Myr

In searching for the cause of a mass extinction we must look for some global change, otherwise many species could migrate to where the environment still suited them. It must also be a sudden change, otherwise many species would adapt through normal evolution.

There is now considerable evidence that the mass extinction around 65 Myr

FIGURE 5.4 Farewell, dinosaurs! (John S Watson)

ago was associated with a comet or asteroid impact. The remains of a large impact crater off the northern Yucatan coast of Mexico – the Chicxulub ("chicksulub") Crater – has been dated at 65.5 Ma and must have been caused by a body about 10 km diameter travelling at about 30 km per second (Figure 5.4). Dust flung into the upper atmosphere caused global cooling. Even though an impact was a factor in this mass extinction, it was certainly not the only one. Some species were already in decline, and there was an increased rate of extinction before the impact, possibly resulting from the climate change that was already under way, notably global cooling and an associated lowering of sea level. Though the impact initiated the decline and disappearance of some species it was just the last straw for others. Many other species were unscathed, including sharks, many reptiles, and the ancestors of the modern mammals.

At the time of the mass extinction 251 Myr ago, plate tectonics had arranged the continents into one large mass called Pangaea, distributed along tropical latitudes. At this time there was a 0.1–0.5 Myr period of extensive volcanic eruptions in what is present day Siberia. This volcanism initially caused a few years global cooling through atmospheric dust, and then, after the dust settled, caused global warming through the increase in atmospheric CO_2. This CO_2 also gave rise to acid rain that made the oceans acidic and lowered their oxygen content, causing huge extinctions among sea creatures. The global mean surface temperature might have risen by up to 5°C. Many land species died and the (smaller) increase in ocean temperatures could only have made matters worse there. The increased ocean temperatures might have caused CH_4 to be released from methane hydrates in the oceans. CH_4 is a powerful greenhouse gas and could have caused up to a further 5°C rise in global mean surface temperature, resulting in yet more extinctions.

The causes of other mass extinctions

The three mass extinctions at about 444 Myr, 360 Myr, and 210 Myr, are associated with climate change, including the onset of ice ages, though the causes of these changes are poorly understood. If, as seems likely, there have also been mass extinctions before 500 Myr, we can only speculate on their causes. As well as the possible causes cited above, there might have been others. For example, there might have been a mass extinction when the oxygen content of the atmosphere began to increase around 2,400 Myr ago – at that time it would have been a poison to most species, a major pollutant. Even today it is poisonous to many prokaryotic anaerobes. Another possibility is a nearby stellar explosion, called a supernova, which would have bathed the Earth in UV radiation, X-rays, and gamma radiation. This might have happened roughly every few hundred million years.

Mass extinctions have probably promoted biological diversity and complexity. It is, however, possible to have too much of a good thing. If asteroid and comet impacts on the Earth had been more common, we might have had sterilization of the Earth, perhaps extinguishing life forever, or, more likely, holding back the emergence of large multicellular organisms.

5.4   THE EFFECT OF THE BIOSPHERE ON THE EARTH'S ATMOSPHERE

Recall that the Earth's atmosphere now is dominated by O_2 and N_2, whereas at the distant time of the last common ancestor, as pointed out above, it is thought to have been dominated by CO_2 and N_2, plus smaller quantities of CO and CH_4, with only a trace of O_2. Water vapor seems always to have been present in significant amounts. It is remarkable but true that if there had never been life on Earth the atmosphere would probably be more like it was at the time of the last common ancestor than it is today. I'll focus on O_2 and CO_2.

Oxygen

In Section 3.3 I pointed out that oxygenic photosynthesis uses H_2O and CO_2 to synthesize sugars, with O_2 released as a by-product. This is by far the dominant process by which O_2 has been released into the atmosphere. Much smaller atmospheric quantities arose from the dissociation of H_2O by solar UV radiation. For a long time there was a balance between the rate at which O_2 was being released and the rate at which it was being removed through its combination with new surface rocks and volcanic gases. Then the balance was lost.

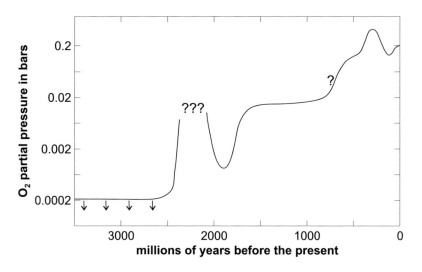

FIGURE 5.5 The build-up of atmospheric O_2 on Earth.

From around 2,400 Myr to about 2,300 Myr there was a rapid increase in atmospheric O_2, from a mere trace to perhaps 1–10% of its present value. Though some of this increase was probably due to a decline in the rate of creation of new surface rocks and volcanic gases, there were probably biogenic contributions, notably the flourishing of oxygenic cyanobacteria, perhaps due to an increase in nutrients in the oceans.

Since then, the geological and biological evidence shows that the amount of O_2 in the atmosphere has broadly increased, though with significant ups and downs (Figure 5.5). For the last 500 Myr or so it has been within 10% of the present amount. As before, a combination of geological and biological factors is believed to have been responsible for the significant variations. One biological factor could have been an increase in the sheer mass of living creatures and their remains, representing a repository of carbon from CO_2, and a reservoir of O_2 in the atmosphere.

It seems beyond reasonable doubt that much of the O_2 in our atmosphere has been put there by the biosphere, at the expense of CO_2.

Carbon dioxide

Atmospheric CO_2 has not only been removed by the formation of organic compounds through photosynthesis. Some is dissolved in the oceans, from where huge amounts have been removed by the formation of minerals called carbonates. Sea creatures have greatly accelerated the process through the formation of shells of $CaCO_3$. When these die the shells settle to the ocean floor. And so it goes on, generation after generation.

Though only 0.037% of the molecules in the atmosphere today are CO_2, it enables oxygenic photosynthesis to take place, without which the surface would be devoid of green plants, algae, certain prokaryotes, and all life that depends on them. No atmospheric CO_2 would mean none dissolved in the oceans, so the oceans would likewise be heavily depleted in life. The trace also makes an essential contribution to the greenhouse effect, without which it would be too cold for life at the Earth's surface.

Global warming

Today, the quantities of most of the constituents of the atmosphere, certainly the dominant constituents, O_2 and N_2, vary appreciably only on long time scales, tens of Myr or longer. This has also been the case far into the past. Such a near perfect balance on a 10 Myr time scale is called quasi-equilibrium. It means that the processes that place these dominant constituents into the atmosphere do so at a rate that very nearly equals the rate at which they are removed from the atmosphere.

Unless you're a recent arrival from beyond the Solar System, you'll know that one atmospheric constituent that is not in quasi-equilibrium is CO_2. In the astonishingly short time since 1700, the proportion of CO_2 molecules has risen from a little under 0.028% to a value of nearly 0.037% in 2000. The rate of increase is itself increasing (Figure 5.6). The value now (2007) is approaching 0.038%.

It is beyond reasonable doubt that this increase is because of an increase in the rate of release of CO_2 into the atmosphere, without a balancing increase in its rate of removal. This has increased the greenhouse effect and is surely the main cause, perhaps the sole cause, of the global warming over the past few decades.

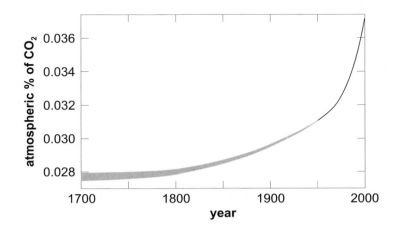

FIGURE 5.6 The rise in atmospheric CO_2 since 1700. The width of the line is a conservative estimate of the uncertainty.

Were it not for the oceans absorbing atmospheric CO_2, the atmospheric increase would be larger, perhaps twice as large. This would have resulted in far larger climate changes, linked to global warming, than those that we are currently experiencing.

There is little doubt that humanity is largely, even solely, responsible for the CO_2 increase. This is because of our increased burning of fossil fuels, and forest clearance, both of which reverse the effect of photosynthesis by using O_2 to convert carbon in living and dead organisms, long dead in the case of fossil fuels, into CO_2. Even if we reduced CO_2 emissions to zero today there would still be substantial climate changes by mid-century as the oceans gradually warm up. This is a legacy of what we have done in the past. All we can do is to reduce our emissions in order to ameliorate the changes.

On this somewhat alarming note, the time has come to consider whether life exists elsewhere, first, in the Solar System – this is the subject of Chapter 6 – then, in the subsequent chapters, beyond the Solar System.

6

Extraterrestrial life in the Solar System?

We can hardly sweep off into outer space without taking a quick look at possible habitats elsewhere in the Solar System. A whole book could easily be written on this alone, and many have been, but my focus in this Chapter is to see what the Solar System teaches us about the possibility of life beyond the Solar System.

For life as we know it – life based on complex carbon compounds and liquid water – we need planets or satellites, and not just any sort. In particular, we need bodies with solid surfaces to provide a stable platform, possibly covered wholly or partially in water, and with a sufficient supply of carbon and the other chemical elements found in living cells. The giant planets can be ruled out – all four are fluid throughout.

We also need suitable temperatures. They must be low enough for complex carbon compounds to survive, which you have seen places an upper limit of about $160°C$ (Section 3.4). At the other extreme, they must be high enough for life to be active, rather than dormant. This requires the water in the cell to be liquid, which, with the aid of dissolved substances needs temperatures above $-18°C$. If temperatures are always lower than this, life will die out, and if they have always been lower than this, life based on liquid water would never have appeared.

Some other conditions might be necessary for the emergence of multicellular life at the surface, such as protection from excessive bombardment, but here I am concerned with life of any sort, even if it is unicellular life with cells of a simple kind, such as prokarya on Earth. To find just one example of life elsewhere in the Solar System would greatly increase the probability of life having emerged beyond the Solar System.

On which bodies in the Solar System, beyond the Earth, are there the right conditions on the right sort of bodies? Have we found any life on any of these bodies? If not, why not?

6.1 THE CLASSICAL HABITABLE ZONE

When considering the possibility of life at, or near, the surface of a planet, we need to focus our attention on what is called the classical habitable zone, HZ. It applies to *all* planetary systems, not just the Solar System. This is the distance

FIGURE 6.1 Goldilocks and the three bowls of porridge. The temperature of the middle bowl is just right. (Graham Read)

from a star in a planetary system, where a planet broadly like the Earth, rocky in its outer composition and with an atmosphere, has its surface warmed by the radiation from its star such that water can be stable as a liquid over at least a substantial proportion of the planet's surface. Any closer, and all the water would vaporize. Any further, and the surface water would be ice everywhere. The HZ is often called the Goldilocks zone – the temperature is just right (Figure 6.1).

The location of the HZ depends on

- the atmospheric model adopted for a planetary atmosphere; and
- the criteria set up for the boundaries of the HZ.

The most important features of an atmospheric model are the composition of the atmosphere and its mass per unit surface area of the planet.

Regarding the boundary criteria, for the inner one I favor the criterion in which a planet at this boundary is just near enough to its star that it lost *all* the liquid water from its surface rapidly via what is called a runaway greenhouse effect. Water is a powerful greenhouse gas (Section 2.3). It has a tendency to cause a runaway because an increase in surface temperature increases the quantity of water vapor in the atmosphere, which further increases the surface temperature, which results in a further increase in water vapor, and so on. If the stellar radiation is sufficiently intense the evaporation will become complete. Not only will the surface then be devoid of liquid water, the surface temperatures will be at least 374°C, well above the 160°C limit for complex carbon compounds, so carbon-liquid water life would be impossible.

At the outer boundary of the HZ the low temperatures ensure that there is little mass of water in the atmosphere. CO_2, among the common volatiles, is then likely to be the controlling substance, and its greenhouse effect is crucial.

The criterion for this boundary that I prefer is the maximum distance from the star at which a cloud free CO_2 atmosphere could maintain a mean surface temperature of $0°C$. This maximum distance occurs with a substantial CO_2 content. To get a feel for how much, an Earth-mass planet at this boundary would need to have about 200 times more CO_2 in its atmosphere than the trace found in ours (0.037%). At greater quantities the surface is cooler because of the stellar radiation scattered back to space by the atmosphere, and at lower quantities the CO_2 greenhouse effect is weaker, so the surface is again cooler.

To establish the corresponding HZ boundaries, a commonly used atmospheric model is by the US astronomer William Kasting and his colleagues. The acid test is where the model plus the criteria place the HZ in the Solar System. If its location fits with what we know about the planets, then we can use the placement in the Solar System to derive the placement of the HZ in other planetary systems, as you will see below. So, what does the acid test reveal?

The classical habitable zone in the Solar System today

The HZ for the Solar System today is shown by the shaded ring in Figure 6.2. You can see that Venus is closer to the Sun than the inner boundary of the HZ. This Earth sized neighbor of the Earth is shrouded in cloud (Figure 6.3), not droplets or ice crystals of water, but mainly droplets of sulphuric acid! Its thick CO_2 atmosphere and surface are extremely dry – there is no liquid water, and precious little water at all. Its thick atmosphere, nearly 100 times the mass of the Earth's

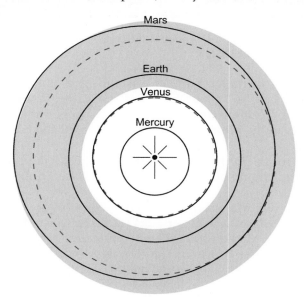

FIGURE 6.2 The classical habitable zone in the Solar System today (shaded ring), and when the Solar System was born, nearly 4,600 Myr ago (dashed circles).

FIGURE 6.3 Cloud shrouded Venus, imaged in the ultraviolet by Pioneer Venus Orbiter. North is at the top. (NASA, P790226)

atmosphere, plays the major part in sustaining a global mean surface temperature (GMST) of 467°C, initially the result the result of a runaway greenhouse effect involving water. This runaway is to be expected of a planet closer to the Sun than the present day inner boundary of the HZ. Venus fits our HZ model.

Then we come to the Earth. This sits firmly in the HZ, and indeed we find water to be stable as a liquid over nearly its entire surface! So this fits our model too.

Our neighbor as we move out from the Sun is Mars (Figure 6.4). You can see that it is at the outer edge of the present day HZ. In fact, liquid water cannot be stable anywhere on the Martian surface – it soon freezes. At first sight this seems at variance with our model. However, Mars is only 51% of the Earth's radius and only 10.7% of the Earth's mass. For several reasons, including the low gravity and the absence of plate tectonics to cycle volatiles out of the crust, the atmosphere of Mars is thin, and though 95% consists of the greenhouse gas CO_2, which gives it rather more CO_2 than there is in the Earth's atmosphere, it only raises the GMST by about 5°C, from –60°C to –55°C. On Earth the greenhouse effect raises

FIGURE 6.4 Mars, a view from the Hubble Space Telescope (in Earth orbit) in August 2003 when it was particularly close to the Earth – just 55.8 million km away. It has a thin atmosphere overlying a solid surface. The south polar cap is in view. It consists of CO_2 ice (dry ice), thought to be underlain by water ice. (NASA/STScI, J Bell, M Wolff)

the GMST by about 33°C, from –18°C to 15°C. About two-thirds of the Earth's greenhouse effect is due to water vapor, the rest to CO_2. On Mars it is too cold for water vapor to be more than the merest trace in the atmosphere, and so the greenhouse effect is due almost entirely to CO_2.

If Mars were a bigger planet then it would have a more massive atmosphere than it does today, with higher surface pressures, and water would surely be stable as a liquid over much of its surface. Its present day location in the HZ would then fit the model. You have seen that the model is based on an atmosphere where the main greenhouse gas is CO_2, a fair assumption at a boundary where the GMST is 0°C, thus depleting the atmosphere in water vapor. Other gases, notably CH_4 can alter the outcome, as noted above.

Mars is simply too small for liquid water to be stable at its surface. We must bear this in mind when we consider planets around other stars.

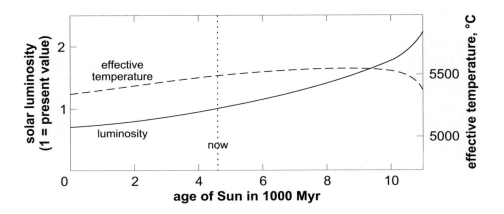

FIGURE 6.5 The increase in luminosity and effective temperature of the Sun throughout its 11,000 Myr main sequence lifetime.

The classical habitable zone in the Solar System at its birth and in the future

Figure 6.2 also shows the HZ in the Solar System at its birth, nearly 4,600 Myr ago. You can see that the HZ was then significantly closer to the Sun than it is today. This was because the Sun was then less luminous. As the Sun evolved, the rate of power released by hydrogen fusion in its core increased, and this caused an increase in the Sun's luminosity, that in turn caused an increase in the surface temperature of a planet. This moves the HZ outwards.

Figure 6.5, based on excellent models of stellar evolution, shows the increase in the Sun's luminosity from its birth to when, at an age of about 11,000 Myr, it starts to evolve rapidly towards becoming a red giant star.

This Figure also shows the Sun's effective temperature (Section 2.2). Up to an age of about 8,000 Myr you can see that the effective temperature increases. Consequently, the radiation from the Sun becomes more dominated by visible radiation at the expense of infrared radiation. This reduces the effect of increasing luminosity, because visible radiation from a star is *less* effective than infrared radiation in heating a planet. However, the increase in luminosity has the larger effect, and so the inner boundary of the HZ does indeed shift outwards as the Sun ages.

As the outward motion continues, there will come a time when the Earth is no longer in the HZ. It will then be uninhabitable. Quite when this will occur is uncertain, with estimates ranging from 800–4,400 Myr into the future, in any case well before the transition of the Sun to a red giant begins. When the HZ has left the Earth behind, our planet will then be too hot for liquid water, and might suffer a runaway greenhouse effect, and come to resemble Venus today.

The classical habitable zone around other stars

Our model has thus yielded reasonable locations for the HZ in the Solar System both today and in the past. To obtain the HZ for other stars, all we need to do is to make adjustments for the star's luminosity and effective temperature compared to those of the Sun. We can then tell whether any of the planets orbiting the star do so in the HZ in its present location, or have done so in the past, or will do so in the future.

Beyond the classical habitable zone

The HZ does not encompass all the locations in a planetary system where potential habitats can be found. It is based on the star as the source of heat for a planetary surface. Planets have warm, even hot interiors, due in part to the internal heat generated when they formed, supplemented by heat from the decay of radioactive isotopes. Even little Mars is thought to have a warm interior.

There is also tidal heating. You are familiar with the ocean tides. These are caused by the gravity of the Moon, and to a lesser extent the gravity of the Sun. The Moon and the Sun also cause tides throughout the Earth's interior, though the solid surface only moves through about half a meter. This, however, corresponds to a periodic flexing of the body of the Earth, and this results in heating – tidal heating. It makes a very small contribution to heating the Earth, but this is not the case for the large satellites of Jupiter – the case of Europa is discussed in Section 6.3.

Thus, even in the outer Solar System, where solar radiation is feeble, there may be regions beneath the surface of a body where water could be liquid.

Tidally or solar heated, where are we most likely to find extraterrestrial life in the Solar System now? Failing that, where might we find fossils of extinct life?

6.2   LIFE ON/IN MARS?

Mars, as I pointed out above, is only 51% of the Earth's radius and only 10.7% of the Earth's mass. In Figure 6.4 the south polar cap is clearly visible at the bottom. Note from the caption that it consists of CO_2 ice, thought to be underlain by water ice. The rest of the image shows dark and light regions, each consisting of different minerals, mainly in the form of sand and dust, particularly in the dark regions. The red tint is due to compounds rich in iron.

In the nineteenth century and up to the 1960s, it was widely believed that the dark areas were vegetation. More spectacularly, it was believed by some astronomers that there were canals on Mars, maintained by intelligent Martians for the distribution of a meagre water supply. The American astronomers Percival Lowell, and his contemporary William Pickering, were the main proponents of this idea. It was based on faint markings at the very limit of what could be seen. Figure 6.6 shows one of Lowell's most spectacular drawings, obtained at the

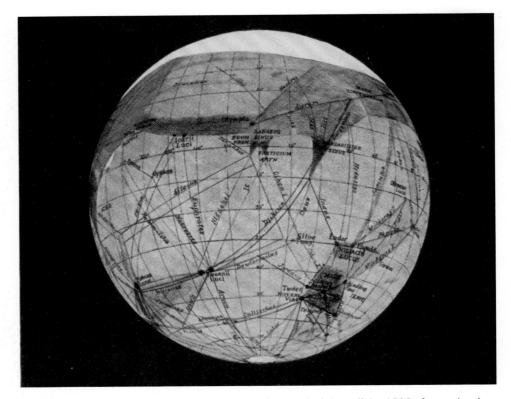

FIGURE 6.6 The "canals" of Mars, drawn by Percival Lowell in 1905, from visual observations through a 0.61 meter aperture telescope at the Lowell Observatory, Flagstaff, Arizona.

Lowell Observatory in Flagstaff Arizona, which he had founded in 1894 for observations of Mars. Unfortunately, many observers could not see the fine linear features that Lowell drew, and we now know that they are constructs of the human mind, struggling to make sense of detail at the very limit of what could be seen.

By the time NASA's Mariner 4 flew by Mars in July 1965, the first spacecraft to visit Mars, observational evidence was heavily against there being canals on Mars. Mariner 4 made measurements of the Martian atmosphere. It was found to be of very small mass, far less than had been thought, and its dryness was confirmed. The belief that there was vegetation on Mars waned. In 1971 it died. In November of that year NASA's Mariner 9 arrived at Mars and went into orbit. There were certainly no canals, and no vegetation.

In 1976, NASA's Viking 1 and Viking 2 went into Martian orbit, and each delivered a lander to the surface. Among several experiments, three were designed to detect evidence of life on Mars. No unambiguously positive evidence was obtained, though more comprehensive, more sensitive measurements would have led to a firmer conclusion, one way or the other. Since the Vikings, many spacecraft have visited Mars and there have been a few more landers. But there have been no measurements specifically designed to find life, and no incidental evidence has emerged. In addition, unlike our atmosphere, the atmosphere of Mars shows no evidence of having been modified by life.

So, is Mars dead? Has it always been dead? Not necessarily. Microbes even a few meters below the surface could easily have escaped detection. But carbon-liquid water life requires liquid water. Is there any evidence for liquid water on Mars today, or in the recent past? Yes, there is. Figure 6.7 shows two images from NASA's orbiting spacecraft Mars Global Surveyor. The image on the left shows a gulley as it appeared in 1999 and 2001, and that on the right its appearance in 2004–5. Did water flow down this gulley between these two dates, and then freeze? Perhaps, though subsequent images from NASA's Mars Reconnaissance Orbiter are consistent with a flow of sand or loose dry dust from bright material uphill. However, the gullies themselves, here and elsewhere on Mars, offer strong evidence that liquid water has flowed on Mars within the last few million years. Although atmospheric pressure almost everywhere on Mars does not allow water to be liquid, and likewise for temperature, conditions beneath the surface could allow water to be stable as a liquid, and this could erupt onto the surface and survive for a short time before freezing.

If, as seems likely, water exists at no great depth in Mars, then life could exist at no great depth too. The deeper we go the more likely there is to be liquid water, so even if there is no life near the surface, there could be life at, say, a few kilometers depth. As in the Earth, this would be microbial life – no Martians in underground passages!

Many other features have been seen on Mars that indicate the flow of liquid water. A selection of these is shown in Figure 6.8. Such features occur mainly on

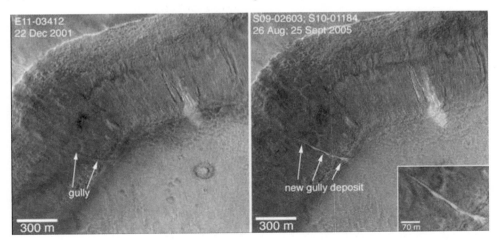

FIGURE 6.7 A gulley on Mars, as it appeared in 1999 and 2001 (left), and 2004–5 (right). These images are from NASA's orbiter, Mars Global Surveyor. (NASA/JPL/MSSS)

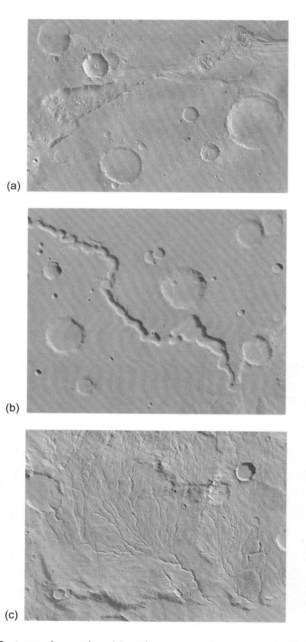

(a)

(b)

(c)

FIGURE 6.8 Features observed on Mars that seem to have been caused by the flow of liquid water. (a) The outflow channel at the head of Simud Vallis. Frame width about 300 km. (NASA/JPL and C J Hamilton, resembles P16893) (b) The fretted channel Nirgal Vallis. Frame width about 160 km. (NASA/JPL) (c) A valley network. Frame width about 130 km. (NASA and C J Hamilton, 63A09)

ancient terrain, older than about 3,000 Myr. This could have been a time when Mars had a more massive atmosphere, and would have been warmer and wetter. With Mars forming nearly 4,600 Myr ago there was plenty of time for life to have evolved, only to become extinct as the small mass of Mars allowed it to lose much of its atmosphere, to the surface, and to space. So, there could be microbial fossils in the terrain preserved from that time. A less optimistic possibility is that Mars was only warm and wet episodically, a result of giant impacts releasing atmospheric constituents from the surface. Such episodes would have been too brief to allow life to start.

Surface life on Mars today can almost certainly be ruled out, but not life deep beneath its surface, nor microbial fossils at shallow depths, even at the surface. Mars deserves further missions tuned to the search for life. In considering planets around other stars we must remember the case of Mars.

6.3 LIFE IN EUROPA?

For extraterrestrial life existing *today* in the Solar System, actually alive *now* as opposed to being fossilized, a better prospect than Mars, indeed the best prospect in the Solar System, is Jupiter's satellite Europa. (Jupiter is ruled out because it is fluid and turbulent throughout.) Figure 6.9 shows the giant planet with Europa's

FIGURE 6.9 Jupiter, with Europa's shadow in front of it (lower left). The Earth would fit about twice across the oval feature (the Great Red Spot) at lower right. (NASA/JPL, University of Arizona, PIA02873)

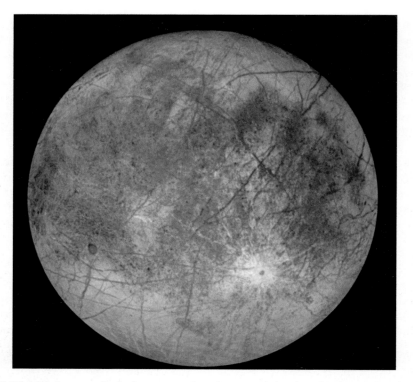

FIGURE 6.10 Europa, displaying a smooth surface consisting largely of water ice. (NASA/JPL., P48040)

shadow as a tiny disc. Figure 6.10 shows Europa itself. If you get the impression of a smooth, ice covered body then that's correct. Europa is covered in water ice, with an altitude range at the surface of only a few hundred meters, remarkably little for a body 1,561 km in radius, 64% that of Mercury. The icy carapace is of uncertain thickness, but it is underlain by a widespread ocean of liquid water. The outer shell of water (ice plus liquid) is about 150 km thick. Beneath it there is a rocky mantle and beneath this an iron rich core.

Jupiter, and its family of satellites, are 5.2 times further from the Sun than is the Earth. The system is thus well outside the HZ. It is certainly not the Sun that has melted ice beneath Europa's surface. It is tidal heating that is responsible. This is the result of Jupiter's gravity, which causes Europa to be slightly distorted. As Europa orbits Jupiter the amount of distortion changes and so does its location with respect to Europa's interior. This varying distortion causes the interior of Europa to be heated to the point where subsurface water ice melts.

The rocky interior is also heated, tidally and radioactively, and it is quite possible that there are structures analogous to the hydrothermal vents found on the ocean floors of the Earth (Section 3.4). If I push this analogy further, then aquatic life forms could exist at such vents, and elsewhere in Europa's oceans.

FIGURE 6.11 An artist's impression of an "aquabot" exploring the oceans of Europa near a hydrothermal vent. (JPL/Caltech/NASA)

Europa is well worth a visit by a lander able to get through the ice into the oceans, and send its findings to a mother ship in orbit around Europa, from where the information would be transmitted to Earth (Figure 6.11).

At present (2007) there is no firm date for a mission to Europa, though there is a European Space Agency-NASA Working Group developing a proposal for an orbiter. I doubt whether there will be any sort of mission to Europa before 2020, alas!

Meanwhile, Europa reminds us that when looking for life beyond the Solar System, we must not confine ourselves to the HZ.

6.4 THE FATE OF LIFE IN THE SOLAR SYSTEM

In Section 6.1 I pointed out that the HZ will continue to move outwards as the Sun's luminosity increases, with the Earth ultimately being left behind and thus becoming uninhabitable some time between 800 Myr and 4,400 Myr in the future. Figure 6.12 is one model of how the HZ moves outwards. It is the variety

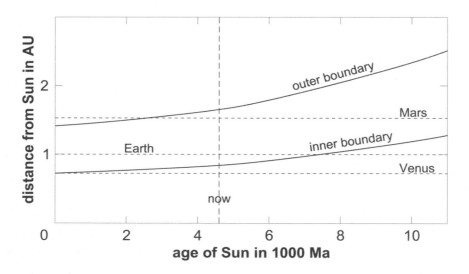

FIGURE 6.12 One model of how the HZ moves outwards during the Sun's main sequence lifetime.

of models of stellar evolution that is largely responsible for the range of times before the demise of life on Earth.

As the HZ moves outwards, the surface of Mars becomes warmer. Substances such as CO_2 and water will be released from the surface and form a much more massive atmosphere, such that liquid water would be stable over its surface. This could happen within a few hundred Myr. Thereafter, Mars will remain in the HZ for thousands of Myr. This is plenty long enough for life to start and evolve, and for any present day subsurface life to spread to the surface. When the Earth becomes too hot for life, Mars would offer a haven, and any life with the necessary technology could get there – as noted earlier, this might not be *Homo sapiens*. Mars could be a home for terrestrials, quite possibly until the end of the Sun's main sequence phase about 6,000 Myr from now.

The end of the main sequence phase is caused by the Sun having fused all of the hydrogen in its core into helium (Section 2.1). The Sun's evolution then speeds up as it starts on its transition to becoming a red giant. The internal events are complex, involving fusion of hydrogen into helium in a shell outside the core, and ultimately the fusion of the helium that now constitutes its core to form carbon. The transition to a fully fledged red giant takes 1,000 Myr or so.

The corresponding exterior events are dramatic. The Sun's luminosity will rise to about 3,000 times its present value. Its surface temperature will have decreased somewhat, giving the Sun a reddish tint, but its radius will have reached around 200 times its present value.

The red giant Sun looms so large over the Earth's horizon in Figure 6.13 that you might fear for the survival of the now barren Earth. Your fears would be well

FIGURE 6.13 An artist's impression of the Sun as a red giant, as seen from the Earth. (Andrew C Stewart)

founded. There is little doubt that Mercury and Venus will be consumed as the Sun expands. If the Sun stayed at its present mass even the Earth might not survive, but red giants emit copious winds, and this reduction in the Sun's mass will cause the Earth's orbit to grow larger, just sufficiently that the Earth retreats in front of the expanding Sun.

During the Sun's transition to a red giant, the HZ surges outwards as the luminosity of the Sun rises. Europa's possible habitability does not at present depend on solar radiation but on tidal heating, and as long as the Sun is insufficiently luminous to melt the icy crust this will remain the case. But as the Sun's luminosity rises the icy crust is likely to melt, and the subsequent loss to space of Europa's water could be rapid, given the low mass of this body – 7.5% that of Mars. Any life could cling on a bit longer beneath what would then be the rocky surface, provided that it was either there already or could colonize the crust rapidly. However, it would be unlikely for life to survive for long. Jupiter itself would survive, its atmosphere expanding in response to the rise in solar luminosity.

Further out, Saturn's Mercury-sized satellite Titan will become warm enough for liquid water, and this might start the prebiotic processes that would lead to the origin of life – if there were enough time. Unfortunately there will only be a few tens of millions of years before the Sun's growing luminosity will boil Titan's surface, and this is unlikely to be long enough for life to emerge. Further out, the prospects for a brief flourishing of life on any of Uranus's satellites seem even bleaker, because of the ever more rapid increase in solar luminosity.

The environs of Neptune and Pluto will probably always be too cold for life to emerge. Even if they become warm enough, then this would be for far too short a time for life to emerge. This is because after the Sun becomes a red giant its luminosity soon drops to about 100 times its present value, and though this is followed by wild upward fluctuations, it will only be a few hundred Myr at most before the Sun ejects a huge mass of sterilizing hot gas. The remnant shrinks to form an Earth sized body called a white dwarf. The accompanying rapid drop in luminosity will result in very low temperatures throughout the Solar System. The white dwarf cools slowly, and fades away.

Roasted, blasted, and then frozen! The only hope for life as we know it would be deep in the giant planets' atmospheres, sustained there as long as heat flows from their interiors. This is to grasp at a straw.

In this manner life in the Solar System will end, a fate that will be, perhaps has been, mimicked in other planetary systems.

7

Habitats beyond the Solar System?

With the essential ground work on the Solar System now covered, the remaining chapters take us much further afield – to planetary systems around other stars. These distant planets, and the possibility of life on them, are the focus of this book. In Section 7.4 I'll describe the requirements for a planet to be habitable. But habitability also depends on the type of star the planet orbits, and on where the star lies in the Galaxy. It is these factors that I focus on in Sections 7.1–7.3, starting with our Galaxy.

7.1   OUR GALAXY

There are many galaxies and there are several different types. We can't, of course, see our Galaxy from outside, but extensive observations from inside show that it has a form that resembles that in Figure 7.1. This shows a face-on view of a galaxy that resembles our own. Our Galaxy is an example of a spiral galaxy, the reason being clear from Figure 7.1. It contains about 200,000 million stars.

In a spiral galaxy the stars, and tenuous interstellar gas and dust, are concentrated into a disc highlighted by spiral arms. In our Galaxy the disc is about 100,000 light years in diameter and most stars are in a thin sheet about 1,000 light years thick – roughly the same ratio of diameter to thickness as a CD. This sheet is called the thin disc. It is enclosed in a thick disc about 4,000 light years thick. The spiral arms are delineated by a high space density of particularly luminous stars and luminous interstellar clouds (the space density is the number of stars per unit volume of space). Elsewhere in the disc the space density of the stars and interstellar clouds is no less – it is just that they are not as bright.

At its center the disc has a roughly spherical bulge called the nuclear bulge, also full of stars and interstellar matter. The bulge is circular in Figure 7.1, though in our Galaxy it is slightly elongated into a bar, about 27,000 light years along its longest dimension. The disc is enveloped in the halo (not visible), a roughly spherical volume in which interstellar matter is particularly tenuous and the space density of stars is low.

FIGURE 7.1 A face-on view of a galaxy that resembles our own, M74, about 30 million light years away. (NASA/ESA/GMOS/HST)

Box 7.1 Astronomical units of distance

In our everyday lives the kilometer (0.621 miles) is a convenient unit of distance. The average adult can easily walk a kilometer (km) in about 15 minutes. It also suits the sizes of planetary bodies – the Earth has an equatorial diameter of 6,378 km. In the Solar System the astronomical unit (AU) is a convenient unit, being the average distance of the Earth from the Sun (149.6 million km). But even the nearest star (Proxima Centauri) is 267,000 AU away. Clearly a bigger unit of distance would be convenient, and one such is the light year.

Note that the light year is a unit of *distance*, and not of time. It is defined as the distance light travels through space in one year. It is equal to 63,239.8 AU, or 9.460536 million million km. Quite a distance! After the Sun, the closest star, Proxima Centauri, is 4.22 light years away.

FIGURE 7.2 Looking towards the center of our Galaxy. The nuclear bulge is peeping above the horizon, and part of the Milky Way stretches above it. (Naoyuki Kurita)

The Sun is located near the edge of a spiral arm, roughly half way from the center of the Galaxy to the edge of the disc. In the night sky at any particular time, from within the disc we can see roughly half of the disc – the other half is below the horizon. A view towards the center of the Galaxy is shown in Figure 7.2 – at the time this photograph was taken the nuclear bulge was peeping above the horizon. The band of light stretching above the bulge consists of very many disc stars, and constitutes (part of) the Milky Way, giving our Galaxy its usual name, the Milky Way Galaxy. Only our side of the nuclear bulge is visible – there is too much interstellar matter to see through the bulge.

Nearly all the exoplanetary systems so far discovered lie within a few hundred light years of the Sun – which is in our cosmic backyard. This is because, broadly speaking, the nearer an exoplanetary system is to us, the easier it is to detect. However, much of the region on our side of the nuclear bulge is accessible to our instruments, or to instruments we will soon have, so there is plenty of accessible space to search.

Beyond our Galaxy there are many more galaxies, by no means all of them spiral. We have available the observable Universe. This is a sphere centered on us with a radius equal to the speed of light times the age of the Universe, 13,700 Myr, so the sphere has a radius of 13,700 million light years. We cannot see any further because there has not yet been enough time for light to reach us. Within the observable Universe there are hundreds of thousands of millions of galaxies, but the chance of detecting extraterrestrial life in galaxies beyond our own is remote. The distances are too large.

I'll therefore concentrate on the Milky Way Galaxy, starting with the stars within it.

7.2   STARS

Types of star – metallicity

Stars differ in several ways that are important to the prospects of finding habitable planets around them.

At the start of their main sequence lifetimes (when hydrogen starts to undergo fusion to helium in their hot cores), they are broadly of the same composition as the Sun – about two-thirds of the mass is hydrogen, about a third is helium, and up to a few percent is all the other 90 chemical elements. This latter fraction is called the metallicity of the star, even though many chemical elements are non-metallic e.g., carbon and oxygen. Most metallicities are in the range 0.05% to 5%. The Sun's metallicity is rather less than 2%.

The importance of metallicity to habitable planets is that if the metallicity is at the low end of the range, then there will be little dust in the circumstellar disc, and so there is unlikely to be a sufficient mass of iron and of the elements that comprise icy and rocky materials, to make much in the way of planets with solid surfaces (possibly partly or wholly covered in oceans). No such planets, no potential habitats. The metallicity also affects the effective temperature and luminosity of the star, though this has little direct bearing on the types of planets that form.

Types of star – birth mass and its consequences

Stars, like children, also differ in their birth mass, ranging from about 100 times the mass of the Sun, down to about 0.08 times the solar mass. The lower the mass, the greater the number of stars. For example, for each star with a birth mass around 10 times that of the Sun, there are roughly 300 stars with a mass about that of the Sun, and roughly 3,000 with a mass about half that of the Sun.

Luminous objects with masses less than 0.08 solar masses exist, but the interior temperatures never become high enough for sustained thermonuclear reactions, so they are not classified as stars. These small, dim objects are called brown dwarfs. They have their highest luminosity at birth, but so low that planets with habitable surfaces are not expected to be present, particularly as the luminosity declines significantly in times short compared with the main sequence lifetime of the Sun.

The mass of a star has a huge effect on its luminosity, effective temperature, and its main sequence lifetime (at the end of which its transition to a giant star sterilizes its planetary system). Table 7.1 shows the birth values. Birth masses of 3.0, 1.0, and 0.5 times the mass of the Sun are shown. The mass declines as the star ages, as a result of stellar winds, but for stars up to many solar masses only a very small proportion of its mass is lost during the main sequence lifetime. The radius, luminosity, and effective temperature rise slowly as the star ages, though the effective temperature does decline towards the end of the main sequence lifetime (the case of the Sun is shown in Figure 6.5).

TABLE 7.1 The dependence on stellar mass of four properties of main sequence stars.

Birth mass (present solar mass = 1.0)	Birth radius (present solar value = 1.0)	Effective temperature at birth (°C)	Luminosity at birth (present solar value = 1.0)	Main sequence lifetime (Myr)
3.0	1.7	10,000	30	500
1.0	0.8	5,300	0.70	11,000
0.5	0.4	3,000	0.03	100,000

Note that the values in columns 2–5 of Table 7.1 are *approximate* – among other things there is a dependence on metallicity. Note also that there is a continuous range of masses, and therefore a continuous range of dependent properties.

The essential points in Table 7.1 are as follows. The less massive a main sequence star

- the smaller its radius;
- the lower its effective temperature;
- the smaller its luminosity (very much so); and
- the longer its main sequence lifetime (hugely so).

The fourth point is related to the third. Higher mass stars have a greater store of hydrogen fuel in their hot cores, but they are so much more luminous that the store is depleted much more quickly. This is because of higher core temperatures, resulting in a large increase in fusion rates. The transition to the giant phase is thus very much sooner. For a star three times the mass of the Sun this occurs after roughly 500 Myr, a main sequence lifetime too short for life to have evolved very far, if at all, if the Earth is any guide (Section 4.1). By contrast, a star with half a solar mass would have all the time in the world.

Spectral class

Main sequence stars are called dwarfs! The term "dwarf" is in comparison with the much larger giant and supergiant stars. Stars are subdivided into what are called spectral classes. Dwarf spectral classes are determined largely by the effective temperature. The complete classification scheme by descending effective temperature is O, B, A, F, G, K, M, plus a few oddities that won't concern us. A useful mnemonic is "O! Be A Fine Girl/Guy – Kiss Me." Why such a puzzling order of letters? The answer is that this lettering system was set up before the link with effective temperature was established – the original sequence was in alphabetical order, with some letters that have now vanished.

These letter labels each cover a range of effective temperatures, and therefore a range of associated properties.

The 3 solar mass star in Table 7.1 has an effective temperature that places it at the hot end of the A class – the cool end is at about 7000°C. It is thus an example of an A dwarf. The solar mass star is towards the hot end of the G class – it is an example of a G dwarf. The Sun is another G dwarf towards the hot end of the G class. The 0.5 solar mass star is towards the hot end of the M class – it is an example of an M dwarf.

Types of star – effective temperature and color

The effective temperature, and therefore the spectral class of a star, determines its tint, just as the tint of a hot metal is determined by its temperature. Stellar tints are not very strong, but they are distinct (Figure 7.3).

O, B, and A dwarfs have a blue tint, weak in O class, weaker in B class, and barely perceptible in A class. F dwarfs have no tint – these are white. G dwarfs, which include the Sun, have a faint yellow tint, and K dwarfs have an orange tint. M dwarfs have a red tint. I emphasize again that the tints are weak, though very striking when differently tinted stars are in the same field of view (Figure 7.3).

Stars that have completed their main sequence lifetimes and become giant stars also have tints determined by their effective temperatures, in just the same way as main sequence stars. For example, at around 3000°C giant stars have a red tint. Giant stars evolve, and at higher effective temperatures they have a pale orange tint, and for brief periods at yet higher temperatures a pale yellow tint. Giants have evolved from main sequence stars up to about 10 solar masses. More massive stars evolve into supergiants, which at various times in their lives have effective temperatures such that the tints cover the same range as main sequence stars.

FIGURE 7.3 A star field where the tints of stars are apparent. This is the open cluster M50, which consists of several hundred stars. It is about 3,000 light years away, and about 20 light years across. (T Credner and S Kohle, Calar Alto Observatory, University of Bonn)

Box 7.2 Why are there no stars with violet or green tints?

The colors of the rainbow are, from inside outwards, violet, blue, green, yellow, orange, and red. Each of these rainbow colors corresponds to a narrow range of wavelengths. The violet range covers wavelengths shorter than those in the blue range; the blue range covers wavelengths shorter than those in the green range, and so on to the red range that covers the longest visible wavelengths. These ranges constitute near pure colors – they are not tints.

Stars emit over a wide range of wavelengths, different quantities of radiation being emitted at each wavelength (Figure 2.10). The tint we perceive is determined by this spectrum and also by the sensitivity of our eyes to different wavelengths. Our eyes are much more sensitive to blue wavelengths than to violet, so even the hottest stars are seen by our eyes as having a bluish tint.

Green tints are not observed. As the effective temperature rises, the spectrum of the star becomes more and more dominated by shorter wavelengths. Consequently the tints go from red, to orange, to yellow, through white (no tint), to blue. A green tint is missing, because even when the star's spectrum peaks at green wavelengths, the fairly strong emission at wavelengths to each side of green quashes a green bias.

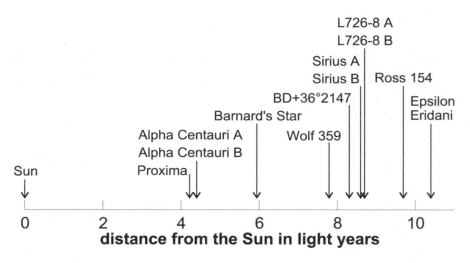

FIGURE 7.4 The distances to the Sun's 12 nearest neighboring stars.

Stars in space

Figure 7.4 shows the distances to the 12 stars nearest to the Sun. Note that there are several multiple star systems, in which the stars orbit each other. In particular, there is the triple system Proxima Centauri/Alpha Centauri A/Alpha Centauri B, and the binary systems Sirius A/B and L726-8A/B. As well as the Sun, only one of the other stars in Figure 7.4 is as yet known to have a planetary system. This is Epsilon Eridani, which, at 10.5 light years, is the closest known planetary system to ours.

If the space density of the stars were independent of the distance from us, then the number of stars lying in a sphere centered on the Sun would be proportional to the volume of the sphere. Out to a few hundred light years this is not far from the truth. Within 30 light years there are about 190 stars, about half of which are in multiple systems, mainly binary systems. Within about 400 light years there are roughly half a million stars.

The stars in our neighborhood are also fairly uniformly distributed in direction. This breaks down beyond several hundred light years, with the number of stars in one direction differing from that in another. This is because we are then encountering the structure of our Galaxy. Clearly, at distances of several thousand light years there will be far more stars in directions that lie in the disc of our Galaxy than in directions that point out of this disc.

Star clusters

Many stars are grouped into clusters. There are open clusters (Figure 7.3) and globular clusters (Figure 7.5). Open clusters consist of a few hundred stars, and are the result of their birth from a giant interstellar cloud, which fragmented.

FIGURE 7.5 The globular cluster M80 which contains hundreds of thousands of stars. It is about 28,000 light years away and about 90 light years across. (STScI – HST)

Open clusters gradually disperse, the one in which the Sun was born dispersed a long time ago. Star formation occurs in the disc of the Galaxy, so this is where open clusters are found.

Globular clusters contain many more stars, typically a million. They formed within our Galaxy soon after it was born. From the age of the oldest stars this is estimated to have been nearly 13,600 Myr ago. Globular clusters therefore consist entirely of very old stars. Unlike the other stars, globular clusters are distributed fairly uniformly throughout the halo and disc of the Galaxy. This means that, because of the huge volume of the halo, it is in the halo that most globular clusters are found.

In Section 7.3 we will consider whether the ancient stars in globular clusters and the young stars in open clusters could have habitable planets.

Stellar end points

After the giant phase, a star sheds a few tens of percent of its mass. The remnant shrinks to become a very hot body about the size of the Earth, but far more massive. You have seen that this is called a white dwarf, which gradually cools. So we have the term "dwarf" applied to two very different types of star, main sequence stars and the far smaller remnant of main sequence stars. This is the case for main sequence stars up to about 10 solar masses.

Main sequence stars with masses *greater* than about 10 solar masses become supergiants. These evolve even more rapidly than giant stars and each ends its life in a supernova explosion. This scatters most of the star's mass into space, leaving behind a tiny remnant only about 10 km in radius, but with a total mass around that of the Sun, and made almost entirely of neutrons. Unsurprisingly, such remnants are called neutron stars. These rotate, and usually radiate beams (at radio, visible, and other wavelengths) in two opposite directions. If one of these beams sweeps across the Earth, rather in the manner of a lighthouse beam, we detect a series of regularly spaced flashes, called pulses. The neutron star is then called a pulsar. Really massive stars, after the supernova explosion, have a remnant that forms a black hole – objects so dense that not even light can escape from them.

Stars in binary systems where the separation is small enough for the evolution of the stars to influence each other, also end their lives dramatically though rather differently. A supernova explosion can occur in some cases. The details will not concern us.

What does concern us is the possible effect of a supernova explosion on life on a planet – more on this in later chapters.

7.3   THE STARS MOST LIKELY TO HAVE HABITABLE PLANETS

Let's bring together various points made in the previous sections to summarize which stars are most likely to have habitable planets. Clearly this will guide us to search the most propitious exoplanetary systems.

Main sequence stars are in a rather stable phase of their lifetime, with nothing changing very quickly nor by very much. By contrast, stars with their main sequence lifetimes behind them are ruled out, with their enormous and relatively rapid changes in luminosity and copious winds. We must also exclude the remnants left after the giant and supergiant phase, particularly the numerous white dwarfs.

Also excluded are stars that do not spend long enough on the main sequence for life to be detectable from afar. For the foreseeable future we will have to scrutinize planets around other stars with instruments on the Earth or in orbit within the Solar System. It is thus not enough for life to have emerged out there; it must also have had some effect on its planet that we could detect.

What main sequence lifetime is long enough? Recall what we know about the evolution of life on Earth (Chapter 5). Of particular significance is the rapid increase in atmospheric O_2 starting about 2,400 Myr ago, due to photosynthesis in the biosphere. You will see in Chapter 11 that an atmosphere rich in O_2 would be detectable from afar. At that time the Sun, which is presently about 4,600 Myr into its main sequence lifetime, was only about 2,200 Myr into it. Therefore, in the absence of anything better, an estimate of the time from the birth of a star to the appearance on one of its planets of some atmospheric (or surface) biological effect detectable from afar, is about 2,000 Myr.

We must therefore exclude O, B, and all but the least massive A dwarfs – their main sequence lifetimes are too short. F dwarfs make it, with main sequence lifetimes from around 3,000 Myr to roughly 10,000 Myr, and of course G, K, and M dwarfs must also be included. What proportion of main sequence stars are O, B, and A class? The numbers of main sequence stars increase strongly from O to M. Table 7.2 shows that in our neighborhood only about 1% are O, B, and A, and so their exclusion robs us of only a tiny proportion of main sequence stars.

TABLE 7.2   The *approximate* proportions of stars in each spectral class out to at least a few hundred light years. All are dwarfs.

O+B+A	F	G	K	M
1%	3%	4%	15%	77%

We must, however, exclude F, G, K, and M dwarfs that are *not yet* 2,000 Myr old. Their exclusion robs us, according to one estimate, of about 10% of them. We must also exclude stars with metallicities so low that they are unlikely to have suitable planets (Section 7.2). One estimate is that 20% of F, G, and K dwarfs have to be ruled out, and perhaps a higher proportion of the exceedingly numerous M dwarfs.

But should M dwarfs be included at all?

It has been argued that *all* M dwarfs should be ruled out. Figure 7.6 shows the HZ of an M dwarf with a mass half that of the Sun, and the HZ of the Sun for comparison. The time axis extends over the whole main sequence lifetime of the Sun. Over this time the HZ of the M dwarf barely moves – this is in accord with its huge main sequence lifetime of about 100,000 Myr. The Universe is only 13,600 Myr old, and therefore all M dwarfs are still in their infancy.

An important feature of the M dwarf's HZ, is that it is close to its star. The gravitational interaction at this range will almost certainly result in any planet in the HZ keeping one hemisphere facing the star all the time, and the other hemisphere facing outwards. This has led some astronomers to argue that the dark side of the atmosphere would act as a cold trap, where the whole of any atmosphere would condense and remain. However, atmospheric models show

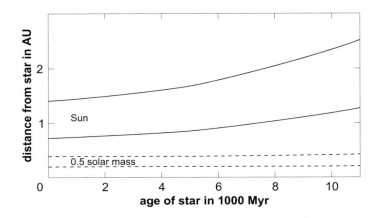

FIGURE 7.6 The HZ of an M dwarf with a mass half that of the Sun, and the HZ of the Sun.

that a CO_2 atmosphere with a pressure of about 10% of the Earth's total atmospheric pressure would prevent this condensation.

Another reason that has been put forward for excluding M dwarfs is that their luminosity is more variable than that of typical F, G, and K dwarfs. Of particular relevance to life is what are called flares.

All stars flare, including the Sun. A flare is a sharp increase in the emission of radiation and charged particles from a restricted region on the photosphere, often associated with star spots (slightly cooler regions on the photosphere). Flares can last up to a few minutes, though a few tens of seconds is typical. A flare consists of a peak of short duration flanked by slow rise and fall. During a flare, the increase in the biologically damaging X-ray and ultraviolet (UV) output can be especially large. However, X-rays would not penetrate a planet's atmosphere, and so it is the UV that is of particular concern.

The UV output from a flare can exceed the normal UV output of the whole star by up to a factor of about 100. Fortunately, the normal UV output of an M dwarf is so feeble that even a factor of 100 increase, would, for a planet with an Earth-like atmosphere, only increase the UV radiation at the surface to a few times the amount the Earth receives from the non-flaring Sun. This would not endanger life.

Though the flare intensity is low, young M dwarfs typically flare far more frequently than the Sun, several times per day in some cases. Fortunately, the frequency declines on a time scale of order 1,000 Myr, and so, even if high frequency were a problem, it might only *delay* the emergence of life on a planetary surface. Life in crustal rocks and in the oceans need not be affected at all.

Another type of variability is in the luminosity of a star due to cool star spots. M stars can have spots proportionately much larger than those on the Sun, so large that they can cause a decrease of a few tens of percent in luminosity for up to several months. However, climate models show that the temperature drop would not be sufficient to eliminate life, not even at the surface of a planet.

Therefore, in the list of stars that are possible hosts of habitable planets, there seem to be no compelling reasons to exclude the abundant M dwarfs. With M dwarfs accounting for about 80% of stars in the solar neighborhood, this increases greatly the number of possible hosts.

Multiple stellar systems

With somewhat over half of the "stars" you see in the sky consisting of two or (rarely) more stars orbiting each other, the question arises of whether these multiple star systems should be excluded. There are two extreme cases to consider.

In the first of these, the stars in a binary system are so close together that the whole HZ lies outside the stellar orbits. In this case the orbiting stars will create a gravitational disturbance in the HZ that will prevent any planets orbiting there. Only at distances well beyond the HZ will stable orbits be possible. As a result about 40% of binary stars have to be excluded – the close binaries.

In the second case, the two (or more) stars in a multiple star system are all at least about 10 AU apart. The HZ of at least one of these stars could then offer stable orbits. It is not well established to what proportion of multiple star systems this possibility applies.

Stars in open clusters are too young for life to have developed. Stars in globular clusters are extremely old and this makes it unlikely that they have suitable planets. This is because when the Universe was very young, at the time the Milky Way Galaxy was born, the proportion of heavy elements was far less than 1%. Therefore the stars formed at that time had very low metallicity, and the circumstellar disc was likewise depleted in the dust that could form suitable planets (Section 7.2).

Conclusion – the galactic statistics

So, when we look up into the sky what proportion of the stars could have planets with life detectable from afar? Indeed, for what proportion of all 200 thousand million stars in the Galaxy is this the case? As well as uncertainties in many of the considerations outlined so far, there is also, on a galactic scale, the possibility of the sterilization of a planetary system by a transient radiation event, such as from a nearby supernova explosion, and the loss of planets through a close encounter with another star.

The answer to the question is thus *very* uncertain. Let's make the optimistic assumption that every star that has survived the cull earlier in this section has at least one planet that is not only habitable, but is inhabited, and furthermore has life that can be detected from afar, regardless of the star's distance from us. In this case, with M stars included, very roughly half of the stars in the Galaxy would have planets with life detectable from afar. If M stars are excluded, then the proportion falls to a few percent. Even at 1% the number of detectable biospheres in the Galaxy is about two thousand million. Unfortunately, we can only see our side of the Galaxy, so the number falls, perhaps by a factor of 10, though this still leaves a few hundred million planets with detectable life.

But what of my optimistic assumption? Even if, as seems likely, the majority of stars have planetary systems, what determines which of them are habitable?

7.4   THE PLANETS MOST LIKELY TO HAVE HABITABLE SURFACES

In this section I set aside tidally heated large satellites of giant planets, like Europa in the Solar System (Section 6.3), and life deep in the crust. This leaves us with life at or near the surface, locations that are most likely to lead to biological effects detectable from afar.

There are three main determinants of whether the surface of a planet is habitable. These are

- the orbit of the planet;
- the composition of the planet; and
- the mass of the planet.

You will know that the orbit of the planet needs to reside in the HZ. By "reside" I mean that the semimajor axis of the planet's orbit must remain in the HZ. A planet with an orbital eccentricity that takes it out of the HZ for part of its orbit will still be habitable.

What about composition? The hydrogen and helium-rich giant planets, as in the Solar System (Section 2.4), are uninhabitable. The problem with such giants, tens to hundreds of times the mass of the Earth, is that they are fluid throughout, with hot interiors. At shallow depths where temperatures are suitable for life, the fluid will be turbulent, disrupting any processes that could lead to life.

As noted at the beginning of Chapter 6, the planet needs a solid surface to provide a stable platform. This can be partly or totally covered with oceans, which could be deep. A habitable planet will thus be dominated by some of the other 90 chemical elements, notably the element iron and the elements that make rocky materials. There must also be small quantities of other materials, notably water, and other substances to form an atmosphere that can stabilize liquid water at and near the surface.

For life at the surface, the mass of the planet needs to be greater than about 30% of the mass of the Earth. This enables it to have enough gravity to prevent escape of much of its atmosphere to space, and sufficient geological activity to prevent the incorporation of much of the atmosphere into the surface by condensation or chemical reactions. In Section 6.1 you saw that Mars, at 11% of the mass of the Earth, indeed has only a very thin atmosphere, and a surface practically devoid of liquid water.

The upper end of the mass range is more difficult to establish. There seems no reason why life could not arise on a rocky-iron planet a hundred times the mass of the Earth. However, the limited availability of iron and rocky materials in the circumstellar dust disc probably imposes a limit of a few Earth masses. Larger masses can occur further from the star, well outside the HZ, where conditions were cool enough for icy materials, notably water, to condense. Masses could easily reach 10–20 Earth masses. In the HZ the ice would melt, resulting in very deep oceans over rocky-iron cores.

To get to the HZ an icy-rocky planet would have to migrate inwards, and then stop migrating in the HZ. Is such migration possible? Yes it is, as you will see in Section 11.3. Migration can also carry giant planets inwards, from beyond the ice line which is where they would have formed (Section 2.4). I have noted that giant planets are not themselves habitable, but large satellites accompanying the giant planet into the HZ could certainly become habitable, irrespective of any tidal heating.

How can we find planets around other stars? How can we tell if they are habitable, even inhabited? What are the results of our searches so far? These questions are addressed in the rest of this book.

8

Searching for exoplanets by obtaining images

I turn now to methods of finding planets beyond the Solar System. These are called exoplanets. Of particular interest are *habitable* exoplanets. In this chapter I shall concentrate on attempts to detect exoplanets by obtaining images. We could then analyze these images to learn about the planet and see whether it is habitable and indeed inhabited, particularly by obtaining the spectrum of the radiation that is either reflected by the planet, or emitted by it.

Imaging is very challenging, more so than the methods that I'll describe in subsequent chapters. In Chapter 9 I'll describe how we find exoplanets through the effect they have on the quantity of radiation we receive from the star they orbit, or from some background star. In Chapter 10 I'll describe how we infer the presence of a planet from its influence on the motion of the star it orbits. We learn less about a planet by these indirect methods, but they currently play a very important role in our search because of their success in discovering exoplanets. In Chapter 11 I describe the exoplanets found so far, and in Chapter 12 the sort we might discover in the future. In Chapter 13 I outline the methods by which we might find life on exoplanets.

8.1 THE CHALLENGE OF OBTAINING IMAGES

Consider a planet orbiting a star about 30 light years away. This is not far compared to our Galaxy's diameter of about 100,000 light years, but it is far enough to be representative of distances in our cosmic neighborhood, and it has come to be used as a standard distance for comparing detection methods. In Section 7.2 you saw that there are about 190 stars within 30 light years of the Sun.

Viewed from 30 light years, the giant planets in the Solar System would not be particularly faint, and would be within the grasp of large Earth-based telescopes (Figure 8.1), the larger the telescope the fainter the objects it can see. Nevertheless the giant planets would at present be undetectable. This is because the Sun is so much brighter than its planets, and, as viewed from a distance of 30 light years is so close to them in direction, that its radiation would drown the far fainter planetary radiation – it's much worse than trying to detect a candle next to a powerful

FIGURE 8.1 The 4.2 meter William Herschel Telescope on La Palma, in the Canary Islands. (Royal Greenwich Observatory)

security lamp. For example, the solar radiation reflected by the Earth at visible wavelengths is about ten thousand million times fainter than the radiation emitted by the Sun. Even Jupiter is about a thousand million times fainter.

At the longer infrared wavelengths the ratios are not so huge, because the infrared radiation emitted by the planet, in accord with its surface temperature, exceeds the infrared solar radiation the planet reflects. Even so, the Sun still outshines the Earth by a factor of about a million and Jupiter by a factor of about 100,000.

These large contrast ratios would not be a problem if a telescope could produce sufficiently sharp images so that the stellar and planetary images were clearly separated. Unfortunately there is a fundamental limit to image sharpness.

A fundamental limit to the sharpness of telescope images

Figure 8.2(a) is a simplified diagram of a reflecting telescope. This is the sort in which the main optical element that collects and focuses the radiation from a distant object is a concave mirror, rather like a shaving mirror. Whereas lenses are still used in cameras, most binoculars, and some small telescopes, the larger telescopes use a concave mirror because compared to a large lens this is much cheaper and will generally have superior performance. The diameter of the main mirror is denoted by D, and is called the telescope aperture. To access the image it has to be intercepted by a small secondary mirror a short distance from the image plane, which deflects the image, for example, to the side, where it can be recorded or even viewed with an eyepiece.

Figure 8.2(b) shows the image of a distant star that would be obtained by the telescope if it had perfect optics and was observing under a perfectly still, clear atmosphere. (Note that this image, on the scale of Figure 8.2(a) would be very much smaller than shown.) The image consists of a central fuzzy disc surrounded by fuzzy rings. Just the innermost ring is shown, which is also the brightest of them, though the others are not a lot fainter. In an *ideal* image a nearby star, being a disc, would be imaged as a tiny disc. For a sufficiently distant star the image would be so small that we would not be able to discern a disc – it would be a point of radiation in the image plane. Even if the image were scrutinized by a magnifying lens (an eyepiece) it would remain as a point.

So, why does our optically perfect telescope, in perfect observing conditions, fail to produce a point image? This is because of a fundamental optical limit called the diffraction limit. A clue to the origin of this limit is in the following two facts. The larger the value of D:

- the smaller the width of the ring pattern in the image plane; and
- the greater the fraction of radiation emitted by the star that is collected.

Putting these two points together, perhaps you can see that one way of accounting for the diffraction limit is that the mirror collects only a (tiny) fraction of the radiation emitted by the star.

Don't dwell too long on this justification! The important point for us is that the larger the value of D the less spread out is the image of the star. Therefore, by increasing D we might hope to make the central disc and the surrounding rings so small that planets are exposed beyond at least the inner few rings. The term resolution is used to describe the spread of the image of a distant object, the smaller the spread the greater the resolution.

For an alien to detect Jupiter from 30 light years at visible wavelengths D needs to be impractically large. Though the contrast ratio between star and planet is smaller at infrared wavelengths, it turns out that the rings are more spread out at these longer wavelengths, so no advantage is gained. We need to do something clever. One clever technique, adaptive optics, is outlined in Section 8.2. Another is coronagraphy.

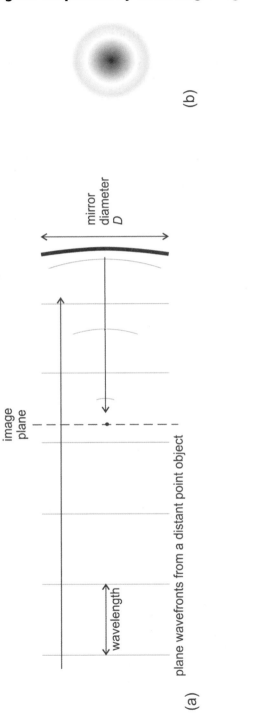

(b)

mirror
diameter
D

image
plane

wavelength

(a)   plane wavefronts from a distant point object

FIGURE 8.2 (a) A simplified diagram of a reflecting telescope that has a main mirror with a diameter *D*. (b) The image of a distant star this telescope would produce if it were optically perfect and was observing under a perfectly still, very clear atmosphere.

Reducing the contrast ratio – coronagraphy

Coronagraphy is named after the Sun's corona – its very faint outer atmosphere. Coronagraphy enhances the visibility of the corona by suppressing the brightness of the Sun's image more than that of the corona. This is achieved by a complex optical layout, the details of which will not concern us.

When applied to a star, it has the effect of reducing the star's image brightness more than the image of the planet. This is possible because the star and the planet lie in slightly different directions. It is then feasible to detect Jupiter from 30 light years, though this would require D to be around 10 meters, and an exposure time of several hours to build up the image on a CCD camera. Detection of the Earth by coronagraphy would require far larger telescopes than we presently possess. Note that a large D also helps to reduce the exposure time, because large D means a large mirror surface to collect radiation. (A CCD "charge coupled device" camera consists of a huge array of tiny electronic detectors.)

There is one further requirement for successful coronagraphy, and that's a very clear, very still atmosphere. But even under the best atmospheric conditions there are still deleterious atmospheric effects. The only way to avoid these completely is to go into space, where coronagraphs are particularly effective. On the ground we have to do the best we can to reduce the effects of our atmosphere, not only for coronagraphy but for any ground-based telescope observations. How can this be done?

8.2   ATMOSPHERIC EFFECTS AND THEIR REDUCTION

The Earth's atmosphere has four deleterious effects on telescope images that prevent us from obtaining images of exoplanets.

First, it *absorbs* radiation from space thus reducing the quantity reaching the ground. Most of the absorption is due to atmospheric gases. The atmosphere is fairly transparent at visible wavelengths, but it is completely opaque in much of the UV, due largely to O_2 and ozone (O_3). It is transparent in the infrared only over certain wavelength ranges; over other ranges it is less transparent, even opaque, due largely to absorption by water vapor in the lower atmosphere. It becomes transparent again at radio wavelengths, except at very long wavelengths where, high in the atmosphere, in what is called the ionosphere, most radiation from space is reflected back to space. Figure 8.3 shows the wavelength regions in the UV, visible, and infrared, in which the Earth's atmosphere is fairly transparent to electromagnetic radiation from space, and those where it is much less transparent, even opaque (100% absorption).

Second, the atmosphere *emits* radiation. Most of the emission arises from the thermal motions of the atmospheric gases. The wavelength range of this thermal emission depends on the temperature of the atmosphere, and for the Earth it is largely in the mid-infrared, particularly in the range 6–100 micrometers with a

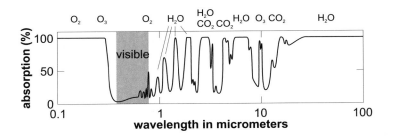

FIGURE 8.3 The wavelength regions showing absorption in the Earth's atmosphere of radiation from space, at UV, visible, and infrared wavelengths.

peak around 10 micrometers. At these wavelengths the atmosphere seems to glow, night and day. This emission makes it more difficult to discern faint celestial objects. By placing infrared telescopes at high altitudes the problems of emission and absorption in the mid-infrared can be reduced, but only by going into space can they be completely overcome.

The third deleterious effect is the *scattering* of radiation. This is where radiation is redirected rather than absorbed. The molecules in the atmosphere are responsible for some of the scattering, the most obvious example being the blue sky, which is the result of solar radiation scattered off atmospheric molecules. It is blue because the intensity scattered increases as wavelength decreases. There is also scattering from aerosols. An aerosol is a suspension of tiny liquid or solid particles, and though aerosols supplied in cans are familiar to you, most aerosols in the atmosphere are of natural origin, and include water droplets, ice crystals, dust, and organic particles. An increasingly troublesome artificial aerosol is contrails (made of ice crystals) from high flying aircraft.

The Sun is such a bright source that scattering of solar radiation from air and aerosols obscures the stars at visible and infrared wavelengths. Scattering of the Moon's radiation also produces obscuration. Other sources of atmospheric illumination at night include ground level artificial lighting, which causes radiation pollution.

As well as degrading images by producing extraneous radiation, scattering also degrades images by scattering back to space some of the radiation from a celestial object, thus reducing its apparent brightness. By siting a telescope away from artificial lighting and at high altitude, these problems can be reduced, but again can only be eliminated by going into space.

The fourth deleterious effect of the Earth's atmosphere is caused by distortion of the waves reaching us from any object in space. This arises from point to point variations in the optical properties of the Earth's atmosphere, and their rapid changes with time due to winds and turbulence. This effect is measured by what is called the atmospheric "seeing". In good seeing the stars shine much more steadily than in poor seeing, when they twinkle a lot – very pretty but not good for astronomy. In poor seeing, in telescopes with D less than about 100 mm the visible image jumps around a lot, whereas in much larger telescopes fine detail is blurred out.

You might despair of being able to do anything at ground level to improve seeing. You would have been right to do so until the last few decades, since when the powerful technique of adaptive optics has been available.

Adaptive optics

Figure 8.4(a) shows the essential features of a typical adaptive optics system. The incoming radiation from a distant star can be thought of as a series of flat planes, wavefronts, moving at the speed of light through space. (The star actually emits radiation over a sphere, but the telescope, even the Solar System, is so small compared to the distance to the star, that the bit we intercept can be regarded as flat.) The flat wavefronts are distorted after passage through the atmosphere. This distortion is removed in two stages. First, the tip-tilt mirror removes the average tilt of the wavefront across the whole area of the mirror. Second, the deformable mirror corrects the smaller scale distortions of the wavefront.

But how does the tip-tilt mirror and the deformable mirror know what to do? The answer is by means of a reference source beyond the atmosphere; in particular a bright star that we know to be a single object and that therefore should produce an image like that in Figure 8.2. Some of the radiation from the reference source is deflected to the wavefront sensor, which evaluates the distortion by comparing the actual wavefront received with the wavefront that would have been received in perfect seeing. The wavefront sensor then feeds a signal to the tip-tilt mirror to remove the average tilt of the wavefront across the whole area of the mirror. It feeds another signal to the deformable mirror that corrects the smaller scale distortions of the wavefront. The ideal outcome is that the wavefronts from the reference source are tilted and flattened to remove atmosphere distortion. These distortions will then be removed from the object of interest, which need not be a single star.

Box 8.1 Artificial reference sources for adaptive optics

The chance of a sufficiently bright star lying sufficiently close to the object of interest is small. Therefore artificial sources produced high in the Earth's atmosphere are usually used. At altitudes of 85–100 km there are sodium atoms that can be made to emit radiation at a yellow wavelength by a laser system that operates at this wavelength. These artificial stars however suffer from the drawback that they are unable to provide the tip-tilt correction. This is because any wavefront tilt impressed on the laser beam when it is traveling upwards is removed when it is traveling downwards, so we learn nothing about the atmospherically induced tip-tilt. Thus, to make the tip-tilt correction a real star has to be used, and this can be the object of interest. This object cannot, however, be used to correct the smaller scale wavefront distortions – this would require us to know its detailed structure, and this is what we are trying to find out! Therefore, to correct these small scale distortions we have to use an artificial star.

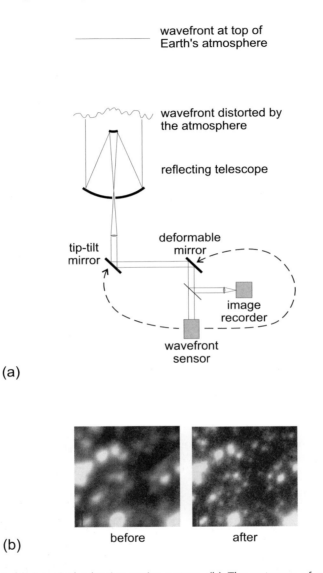

wavefront at top of
Earth's atmosphere

wavefront distorted by
the atmosphere

reflecting telescope

deformable
mirror

tip-tilt
mirror

image
recorder

wavefront
sensor

(a)

(b)   before   after

FIGURE 8.4 (a) A typical adaptive optics system. (b) The outcome of adaptive optics applied to a particular star field.

The correction so far described is for the atmosphere at one particular instant. But the atmosphere is not frozen in place. It evolves through winds and turbulence, requiring that the tip-tilt mirror and the deformable mirror are reconfigured several hundred times per second.

Figure 8.4(b) show an example of an outcome of applying the overall correction. Significant sharpening of the image of the star field has been achieved.

The adaptive optics correction is never perfect, even for the reference source. This is because it takes time to accumulate enough radiation from the reference source to enable the wavefront distortion to be evaluated. During this time the atmospheric distortion changes, even during the short times that are possible with a bright reference source. Also, the deformable mirror is made up of a finite number of elements, perhaps a few tens, and so does not correct the wavefront on a scale smaller than the elements. Moreover, the correction gets worse with increasing angular separation between sources and reference. This is because the radiation from the two sources takes increasingly different paths through the atmosphere to reach the telescope mirror, and thus arrive at the mirror increasingly differently distorted.

Consequently, adaptive optics corrections remove the effects of seeing over small fields of view – at visible wavelengths perhaps a hundred or so times smaller than the full Moon. Performance is better, but not hugely so, at infrared wavelengths. What can be achieved when adaptive optics is used on the largest telescopes that we have? What can be achieved with large telescopes in space, where adaptive optics is unnecessary?

8.3   LARGE OPTICAL TELESCOPES, GROUND-BASED AND IN SPACE

An optical telescope is one that works at either visible, UV or infrared wavelengths. The optics is broadly similar across this wide wavelength range, though a particular telescope and its detectors will be optimized to work over just a part of the range. In Chapter 14 you will meet radiotelescopes, which have a substantially different structure.

Existing ground-based optical telescopes

Currently, the largest ground-based optical telescopes have main mirrors 8–11 meters in diameter. Prominent among these are the four 8.2 meter telescopes that constitute the European Southern Observatory's VLT group at an altitude of 2640 meters on Cerro Paranal in Chile. "VLT" stands for "Very Large Telescope"! Another example is the two 9.8 meter telescopes in the Keck group at an altitude of 4150 meters on Mauna Kea in Hawaii. Let's consider their potential for obtaining images of exoplanets.

At the VLT, it is possible to image large exoplanets orbiting some distance from relatively faint stars. It can achieve this at visible wavelengths, and in the near infrared (which extends to wavelengths of a few micrometers). With any one of the VLT telescopes, adaptive optics and coronagraphy has to be used. But these alone are not enough. In addition, a technique is employed in which one image is subtracted from another, and the images are chosen so that this suppresses the residual radiation from the star that would otherwise obscure the planet. In this way, a Jupiter-twin could be imaged at a range of 15 light years in one night's exposure. A Jupiter-twin, as its name suggests, is a planet the size of

FIGURE 8.5 The European Southern Observatory's VLT group of telescopes at an altitude of 2640 meters on Cerro Paranal in Chile. The four largest each has an aperture of 8.2 meters. (ESO, PR photo 43a/99)

Jupiter, and, like Jupiter, 5 AU from a solar type star (a G dwarf) about the age of the Sun.

Age is important because giant planets are very hot when they are young, and therefore very much brighter and easier to image, particularly in the near infrared. Indeed, one of the first images of a candidate exoplanet was obtained in the near infrared, the planet orbiting a very young star. The image was obtained by the VLT telescope called Yepun. This image is shown in Figure 8.6. The star and planet are only about 1 Myr old, and consequently the planet's outer regions are hot, about 1700°C. The planet is nearly twice the diameter of Jupiter, but its mass is poorly constrained at present and it could even exceed 13 times the mass of Jupiter, in which case it is likely to be a brown dwarf (Section 7.2). The star, GQ Lupi, is a K dwarf with a mass about 70% that of the Sun (star names are a subject for Chapter11). It is about 450 light years away.

On one of the Keck telescopes (Keck II) there is also a camera that operates in

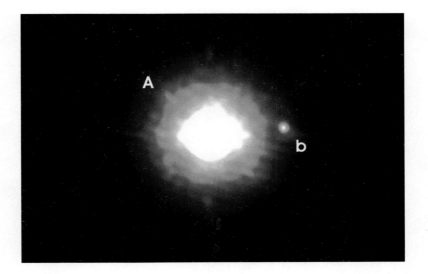

FIGURE 8.6 The very young star GQ Lupi (A) and its hot planetary companion (b), about 100 AU from the star and nearly twice the diameter of Jupiter. The image of the star is overexposed. (ESO/VLT, PR photo 10a/05)

the near infrared. This camera has the capability to image giant planets, 1–10 times the mass of Jupiter, and less than about 60 Myr old. To achieve this, the Keck adaptive optics system has to be used, plus a technique called unsharp masking that enhances point like objects (such as planets) at the expense of diffuse scattered radiation. A survey has been carried out for "young Jupiters", but none has yet been discovered.

As well as obtaining images across a broad range of wavelengths, searches can also be carried out over narrower ranges. One such search, completed in 2007, used one of the VLT 8.2 meter telescopes and the 6.5 meter Multiple Mirror Telescope (MMT) on Mount Hopkins in Arizona. The detectors were particularly sensitive to methane, a gas expected in the atmospheres of giant planets. Fifty four nearby, young stars were observed. No planets were detected. The significance of this null result is a subject for Chapter 11.

There are other large ground-based telescopes that are being used to search for exoplanets, but my intention here is to give you important examples of such telescopes, not a comprehensive review.

Future ground-based optical telescopes

We could of course image fainter planets, or planets closer to their star, perhaps even Earth-twins, if the mirror were considerably larger. An Earth-twin is a planet the size of the Earth, and, like the Earth, 1 AU from a solar type star about the same age as the Sun. Whereas US proposals for extremely large telescopes (ELTs) center on 20–30 meter diameters, Europe is thinking bigger. Current proposals

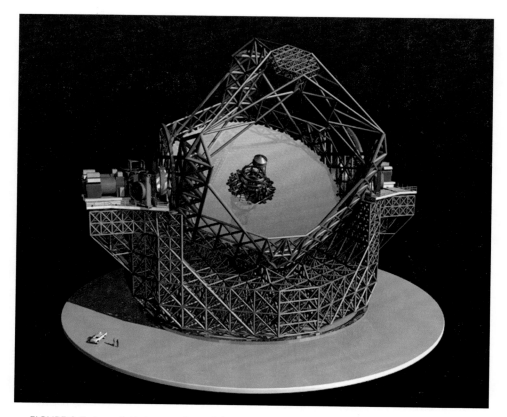

FIGURE 8.7 An artist's impression of the European Southern Observatory proposal for a European Extremely Large Telescope (E-ELT), with an aperture of 42 meters. (ESA)

for the European ELT (E-ELT) envisage a telescope with an aperture of 42 meters. The mirror would consist of five units, working as one. The estimated cost is €750 million, and the completion date would be 2017.

With adaptive optics, E-ELT should be capable of imaging a Jupiter-twin up to hundreds of light years away, but probably not an Earth-twin, though planets a few times the mass of the Earth might be detectable out to a few tens of light years. In a few cases E-ELT might allow us to investigate any such "large Earths" spectroscopically, to see if they were capable of supporting life, and indeed to see whether biospheres were present. How spectra are used to discover whether an exoplanet is habitable/inhabited will be described in Chapter 13.

Telescopes in space

Space has the advantage of avoiding all the problems posed by the Earth's atmosphere, which can only be partially compensated at ground level. The best known space telescope is probably NASA's Hubble Space Telescope (Figure 8.8),

FIGURE 8.8 NASA's Hubble Space Telescope, launched in 1990, and now showing signs of old age. Unless its orbit is boosted it will re-enter the Earth's atmosphere in 2010 or soon after. It has a 2.4 meter diameter mirror. (NASA)

from which many fine images have been obtained, including some of those in this book. But with a 2.4 meter diameter mirror its exoplanet imaging capabilities are severely limited, though it has imaged GQ Lupi and its possible planet. In 2013 its successor is scheduled to be launched, the James Webb Space Telescope (formerly the Next Generation Space Telescope). With a mirror 6.5 meters in diameter, and a coronagraph, it will be able to image planets in the infrared. For example, it would be able to see a Jupiter-twin within about 25 light years with a three hour exposure, and more luminous (younger and/or larger) planets further out.

Earth-twins will be beyond the capabilities of the James Webb Space Telescope (unless they are closer to us than is likely). They would not be beyond NASA's Terrestrial Planet Finder TPF-C. TPF-C would consist of a single telescope with a mirror about 8 meters in diameter, fitted with a coronagraph (hence the C) and other devices to reduce unwanted radiation. Design studies are under way, but as yet (mid-2007) there is no funding. The earliest launch date would be 2014.

There is another design under consideration for the Terrestrial Planet Finder, called TPF-I, with a comparable potential to TPF-C for planet imaging. This is likewise unfunded as yet, and the earliest launch date would be 2020. TPF-I would be what is called an interferometer, hence the I. Before I tell you what an interferometer is, there is one more space telescope that I want to tell you about – Spitzer.

Spitzer was launched in 2003, into an orbit around the Sun that trails behind the Earth. It is a small telescope, with a mirror just 0.85 meters in diameter. It is however the biggest space telescope making observation across much of the infrared spectrum. It obtains images and spectra, and it also measures brightness (photometry). It cannot obtain an image of an exoplanet, but it can, in suitable circumstances, measure its surface temperature. It has achieved this for the planet of the star HD149826. This giant planet has an orbit presented almost edge-on to us, and consequently we can detect the difference in the radiation received from the system when the planet moves from being in front of the star to being behind it. This can be done most readily when the planet is so hot that it radiates a lot of infrared radiation. Spitzer has detected the infrared radiation from the planet of HD149826, and found it to be about 2000°C, the hottest known exoplanet.

Another star, HD189733, has a giant planet that also passes behind its star. The mean surface temperature, measured by Spitzer, is about 1000°C, but observations over 33 hours have enabled Spitzer to obtain a crude temperature map of the planet, which shows a 230°C temperature difference between the day (star-facing) side and the night side, the day side being the hotter. Spitzer was also able to discern a broad hot spot on the day side.

There are plans for much bigger infrared telescopes in space, which would be able to isolate the infrared radiation even from Earth sized planets, but these are interferometers. What is an interferometer?

8.4   INTERFEROMETERS

The basic idea of an interferometer is to combine the output of two or more telescopes so that the detail in the image is much finer than each telescope alone could achieve, thus facilitating the isolation of the planetary radiation from the stellar radiation.

The simplest interferometer consists of two telescopes, as shown in Figure 8.9. It is essential that the radiation gathered by each aperture is mixed with the radiation gathered by the other aperture *before* any of this radiation reaches the image plane, where some imaging device is placed, such as a photographic film, or, almost always these days, a CCD. So we must combine radiation and then form the image. When beams of radiation interact in this way they are said to interfere with each other, and the device that creates this interference is called an interferometer, in this case a two telescope interferometer.

The image formed in this simple interferometer bears information about the form of an object, not just with the resolution of an aperture D, but also with the much greater resolution of a single aperture with the much larger size S (for

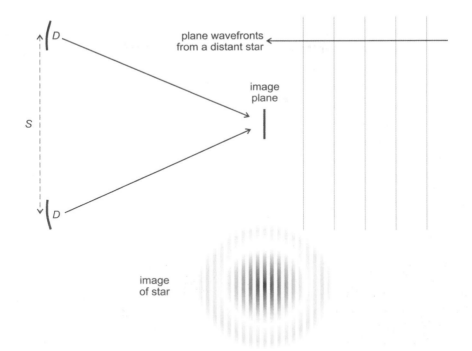

FIGURE 8.9 An interferometer consisting of two telescopes, each with an aperture of diameter D, separated by a distance S. The beams from each aperture mix just before they reach the detector.

separation). The extra information is confined to one direction across an object, the direction of S, and is very limited. Compare the stellar image in Figure 8.9 with that in Figure 8.2(b). The ring pattern in each case is produced by a single mirror. The lines crossing the pattern in Figure 8.9 is the result of having two mirrors. Though I'll not go into details, it's possible to obtain the diameter of the star from a two telescope interferometer

In order to achieve a conventional image with the resolution corresponding to a single mirror with a diameter S, we would have to place many mirrors with a diameter D so that overall they covered an area equal to that of the larger mirror, and then combine all their outputs interferometrically. You have probably spotted that this is no different from using a single large mirror!

However, we can do better by means of an array of 3, 4, 5 ... telescopes. We can do better still by moving these telescopes around and by obtaining an image in each configuration. We will not get an image like that from a single huge aperture, but we will get some two dimensional information at the same resolution as we would have obtained from it. Thus, we could detect the presence of a planet around the star as a patch of light, though other details would be unresolved.

You might ask why, if with an aperture D we can get information appropriate to an aperture S, we don't reduce D to something really small, like 10 mm? Indeed, this would not result in loss of detail, but it would result in a greatly increased set of exposure times. We need to accumulate photons to build up the effect on our detector. Therefore, D needs to be a compromise between being too small to collect radiation at a reasonable rate and too large so that the cost benefit of small apertures is lost.

Interferometers in space

In the previous section you saw that TPF-I will be an interferometer. It will be a multi-aperture infrared interferometer capable of yielding (fringy) images of exoplanets, including Earth-twins.

The European Space Agency (ESA) is also considering a space-based interferometer with a similar capability. This is called Darwin, and Figure 8.10

FIGURE 8.10 An artist's impression of a possible design of ESA's proposed interferometer, Darwin, in a version using six mirrors, each about two meters in diameter, with variable spacing. Note that the telescopes and other components are not connected, but are moved and kept on station by small jets. (ESA, 2002)

shows one possible design. The currently favored version consists of four or five telescopes each with a mirror about two meters in diameter. They lie on the circumference of a circle, will move around this circumference, and move in and out. The diameter of the circle can be varied over the range 7–500 meters. The radiation is fed to a central hub where an image can be reconstructed. Darwin would operate in the mid-infrared, at wavelengths of 6–20 micrometers. This is preferred over the visible range because the contrast ratio between star and planet is smaller (Section 8.1), and it is an informative spectral region about life on an exoplanet (Chapter 13). Note that the various components in Figure 8.10 are not connected, but are flying in formation.

Note that the Earth's atmosphere precludes ground-based interferometry in the mid-infrared. Not only does the atmosphere emit strongly at these wavelengths, there are also absorption bands that block our view of space over much of the infrared (Section 8.2). It is thus greatly advantageous to place an infrared interferometer in space. Space has the further advantage, perhaps surprisingly, that it is easier to maintain the interferometer configuration than on the ground, which vibrates. A configuration either needs to be maintained to a small fraction of the wavelength of radiation used, or we need to measure where each telescope is to this accuracy, and use optical delay lines to correct for positioning errors. With Darwin operating at wavelengths around 10 micrometers you can see that this is a daunting requirement. Darwin would meet this through laser monitoring of the telescope positions and a set of jets on each telescope.

Darwin would make use of what is called nulling interferometry. By adjusting the path differences from each telescope to the hub it is possible for the waves from an object in a direction perpendicular to the mirror array to cancel or be nulled. A star in this direction would be greatly suppressed (not entirely so, because of the finite size of the star). A two dimensional image is built by repositioning the telescopes radially and around a circle a sufficient number of times to obtain the required detail. An image might look something like that in Figure 8.11. This is a simulation of what the Solar System would look like from afar – the three bright objects are Venus, Earth, and Mars. The other structures are spurious, arising ultimately from the incomplete filling of the aperture covered by the telescope separations.

One or both of Darwin and TPF-I could be in space in the 2020s, and either of them would be able to discover Earth-twins out to several tens of light years with a few hours exposure, and with several days or weeks of exposure to investigate them for life (Chapter 13).

ESA has plans for small space missions to test formation flying (as in Figure 8.10). Nulling interferometry is in operation using Keck I and Keck II as an interferometer. At the VLT, by about 2009, the four large telescopes will perform nulling interferometry in a project called GENIE (Ground-based European Nulling Interferometer Experiment). These interferometers pave the way for TPF-I and Darwin, but have some limited planet detecting capability themselves. However, at present, the direct detection of exoplanets is rare from space and

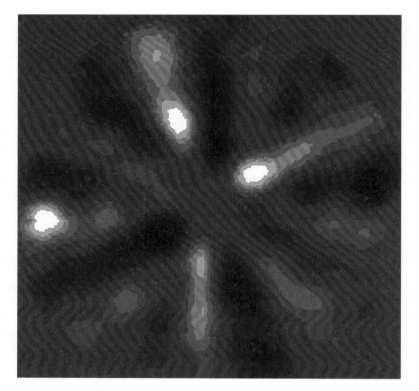

FIGURE 8.11 A simulated image from a five telescope version of Darwin, using nulling interferometry, of the Solar System from afar. The three bright blobs are Venus, the Earth, and Mars. (Courtesy of Bertrand Mennesson)

from the ground. Therefore, we have to rely on indirect methods to discover and investigate the very great majority. These methods are described in the next Chapter.

8.5   CIRCUMSTELLAR DEBRIS DISCS

There is a feature of a planetary system that can be observed far more readily than an exoplanet orbiting its star, and that's a debris disc. This is solid material of the sort expected to be present during and after the formation of planets and smaller bodies such as asteroids and comets. It thus provides indirect evidence that planets might be present. It is the dust component of a debris disc that is observed. This is because a given mass of dust has a much bigger surface area than the same mass contained within objects the size of boulders upwards.

In most cases debris discs are detected at infrared wavelengths from the radiation that the dust emits, the dust having been warmed by the absorption of radiation from the star it encircles. Because the disc is so large it easily outshines the star at such wavelengths. Indeed the first debris discs to be discovered, in the early 1980s, were not imaged but detected as an apparent infrared excess from the star. Later, the discs were imaged, and found to be, well, disc shaped, usually with central holes where it is assumed planets reside. Distortions in some discs provide further evidence for planets.

Over 1,000 stars are known to have debris discs. These have been discovered by ground-based and space-based telescopes. In a few cases discs have been imaged at visible and near infrared wavelengths, through the stellar radiation that they *scatter*. This requires the light from the central star to be blocked. One such example is shown in Figure 8.12, from the Hubble Space Telescope using a camera fitted with a coronagraph. This is an edge-on view of the disc around the star AB Microscopii. The position of the star is marked at the center of the coronagraphs occulting disc, which obscures the disc out to 11.5 AU. For a debris disc to be as substantial as this, the star has to be very young, before any planetary formation is complete. There is a hole in the center of the disc, a few AU across, perhaps due to the formation of planetesimals and embryos (Section 2.4).

AB Microscopii is indeed very young, only about 12 Myr old. Most stars with substantial debris discs are young. This is because the debris gradually gets mopped up until its rate of disappearance is matched by the rate of creation of fresh dust, for example by collisions between small bodies that are the analogs of Edgeworth–Kuiper belt objects in our Solar System (Section 2.3). A debris disc reaches this equilibrium after at most a few tens of Myr. The Solar System is 4,600 Myr old so it will not surprise you to learn that its debris disc is very tenuous. It is very faint in our night skies, but can be seen in very clear skies as a faint band of light stretching along the plane of the planetary orbits. It is brightest towards the Sun, so is best seen not long after sunset or shortly before sunrise. It is called the zodiacal light.

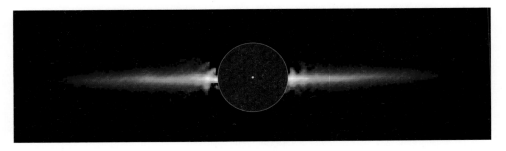

FIGURE 8.12 The debris disc encircling the young M dwarf AB Microscopii, 33 light years away. This image was obtained by the Hubble Space Telescope with the aid of a coronagraph. The position of the star is marked at the center of the coronagraph's occulting disc, which obscures the disc out to 11.5 AU. (NASA/ESA, J Graham (University of California, Berkeley))

Though debris discs indicate that planets might be present, and in some cases almost certainly so, they will not be accredited with planets unless one or more planets are detected.

So, what are the techniques that have been particularly successful in detecting exoplamets? These are included in the next two chapters.

9

Searching for exoplanets by stellar photometry

Astronomers, other scientists, and the public at large would very much like to see images of exoplanets, particularly habitable exoplanets. Such images could then be scrutinized for life and landscapes. Even seeing a planet as a single dot would be an enormous advance in our search for life, because we could then analyze the infrared radiation it emits or the light it reflects, as you will see in Chapter 13. However, in the previous chapter I indicated that, with rare exceptions, this cannot yet be achieved, and not at all in the case of habitable planets – these will be more Earth sized than Jupiter sized and thus pose a much greater challenge to detection methods.

In this chapter I shall concentrate on those indirect methods where we infer the presence of a planet from its influence on the quantity of radiation we receive from this star or from some background star – such measurements are called photometry. In the next chapter I'll concentrate on indirect methods that rely on the effect of a planet on the motion of the star it orbits.

We learn less about a planet through indirect methods, but they have discovered almost all the exoplanets currently known.

9.1   TRANSIT PHOTOMETRY

If the orbit of a planet is presented to us sufficiently close to edgewise, then, once per orbit, for a small fraction of its orbital period, the planet will pass between us and the star – the planet will be observed in transit. The planet emits negligible radiation at visible wavelengths, so the visible radiation we receive from the star during the transit is reduced. Even at infrared wavelengths, where the planet emits radiation, the brightness of the planet is far less than that of the area of the star it obscures, and so the transit is likewise detectable.

If the surface of the star has uniform brightness, and if the whole of the planet passes across the stellar disc, then the *apparent* brightness of the star will drop as shown in Figure 9.1. The depth of the decrease in apparent brightness depends on the cross sectional area of the planet, A_p, compared to that of the star A_{star}. The fractional decrease is simply the ratio of these areas, A_p/A_{star}. For example, if

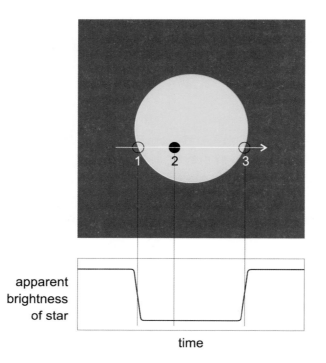

FIGURE 9.1 A planetary transit, and the associated light curve (neglecting limb-darkening of the star).

the planet were so huge and the star so small that the cross sectional area of the planet were a quarter that of the star, the fractional decrease in apparent brightness of the star would be 0.25 i.e., 25%. More realistically the values are about 1% or less.

The one huge advantage of transit photometry is that it gives us the radius of the planet, provided that we know the radius of the star. The stellar radius can be estimated from the star's luminosity and its effective temperature (the details will not concern us). The radius of an exoplanet cannot be derived from the remaining techniques to be described in Chapters 9 and 10, but they do give us the mass of the planet, which we cannot derive from transit photometry. Thus, if a planet is observed in transit and *also* by a method that gives us its mass, the planet's mean density can be calculated (equal to mass/volume), which places a constraint on its composition. This is because different materials have different densities.

As well as the radius of the planet (as a fraction of the radius of its star) the interval between the dips in apparent stellar brightness gives us the orbital period of the planet, and from an estimate of the star's mass the orbital semimajor axis of the planet can then be calculated.

In rare cases we can learn more. When the planet passes behind its star the reduction in infrared radiation received from star plus planet can be combined with the radius of the planet to provide an estimate of the temperature of the zones in the atmosphere or at the surface of the planet that are emitting infrared radiation to space. In other cases it has been possible to detect the radiation from specific elements in the planet's atmosphere, such as sodium.

Complications

So far, the assumption has been made that the star's surface is of uniform brightness. In fact, stars dim slightly towards the limb (the outer rim of the disc as presented to us). This is called limb-darkening. If the transit is near the center of the stellar disc then at mid-transit the decrease in apparent brightness of the star will be slightly greater than A_p/A_{star}. If the transit is near the limb but the planet still fully overlaps the star's disc, the decrease will be slightly smaller than A_p/A_{star}. However, the shape of the light curve is different in each case. Therefore, if we could obtain sufficiently precise curves we could estimate the limb-darkening and thus obtain an improved estimate of A_p/A_{star}.

The limb is darker because the radiation we receive from it is predominantly from the uppermost regions of the photosphere, which are cooler than deeper down. At the center of the disc our line of sight through these cooler regions is less than at the limb.

Another complication is when the decrease in apparent brightness of a star is due to the grazing transit of a dim companion star i.e., when the companion appears to nip the edge of the star's limb. This could be confused with a non-grazing planetary transit, though the light curves differ, and so we could distinguish these cases. Another problem arises if the star undergoing a transit (by a planet or a star) has its light contaminated by a foreground star near the line of sight or by a close companion star. Care has to be taken to detect such contamination and eliminate its effects, the main one being a reduction in the apparent decrease in brightness during transit, leading in turn to an under-estimate of A_p/A_{star}.

What sort of planets can we detect?

If Jupiter were to be observed to transit the Sun then the apparent decrease in solar brightness would be about 0.01 (1%). This is because the radius of Jupiter is about a tenth that of the Sun, and with area proportional to the square of the radius, A_J/A_{Sun} is about 0.1 × 0.1 i.e., 0.01 or 10 parts in 1,000. To obtain useful accuracy of Jupiter's radius, the radiation from the Sun would have to be measured with a precision of a few parts in 1,000. Such photometric precision is readily achievable with ground-based telescopes.

However, if the *Earth* were to be observed to transit the Sun then the apparent decrease in solar brightness would be only about 0.000 08, which is 0.008%, or 8 parts in 100,000. In this case the photometric precision would need to be two or three parts in 100,000. Unfortunately, the fluctuation in transparency of the Earth's atmosphere limits precision to about 1 part in 10,000, and so the detection of Earth-size planets around solar-type stars will have to be attempted in space. Only for M dwarfs less than about one tenth the radius of the Sun, would the detection of Earth-size planets from the ground be feasible.

The ultimate limit to the detection of planets by transit photometry is imposed by variations in stellar brightness that mimic the effect of a transit. There is the possibility of a dip in the star's luminosity for the few hours that are typical of a transit. If only a *single* dimming is observed then for a solar type star the limit is a planet with a radius greater than about half that of the Earth in a 1 AU orbit (like the Earth). In a smaller orbit smaller planets could be detected because the orbital period is smaller, and solar variability on the shorter time scale of such transits is less. For M dwarfs, planets down to about a third of the Earth's radius (Mercury size) could be detected out to a few tenths of an AU from the star. If several dimmings are observed then this helps distinguish a transit from stellar variability, and this can push the planetary size limits downwards.

Another approach is to detect color variations. The limb is not only cooler, but as a consequence is redder than the rest of the disc. In a transit the starlight will therefore be slightly less red when the planet obscures the limb, than when it is more central. Any color variations due to stellar variability would be different. This approach would however require an order of magnitude improvement in photometric precision.

A potentially fruitful approach to the detection of Earth-mass planets is by the accurate timing of many transits of a giant planet. If accuracies of a few seconds can be achieved then any periodic variations of the intervals between transits would reveal the presence of the gravitational tugging of another body, even as small as the mass of the Earth. If the mass of the giant planet is known, the orbit and mass of the perturber could be calculated, but not its size.

What is the expected "strike rate"?

If, as is surely the case, the orbits of exoplanets have random inclinations with respect to the plane of the sky, then we can estimate the proportion of exoplanetary systems that should have transits. Clearly, the closer the planet is to the star, and the larger the star, the greater the proportion. For planets only about 0.05 AU from a star with the solar radius, the proportion is about 10%. At about 5 AU (Jupiter distance) from a solar radius star the proportion has fallen to only about 0.1%. Moreover, whereas at 0.05 AU the transits are spaced by a few days, at 5 AU they are spaced by about ten years.

On the other hand, solar type stars are bright enough for transits to be detected out to many thousands of light years, which provide us with plenty of potential targets.

Search programs

There are at present a few tens of search programs for planetary transits, many of them dating back just a few years. Already the trickle of discoveries has increased

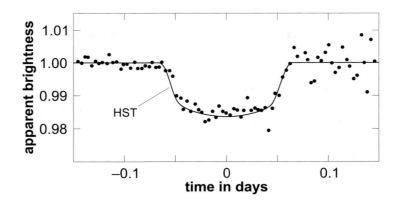

FIGURE 9.2 The first detection of an exoplanet by transit photometry (the dots). Transits were detected in 1999 by means of the 0.1 meter aperture telescope of the STARE (STellar Astrophysics and Research on Exoplanets) project at Boulder, Colorado. The curve labelled HST (Hubble Space Telescope) was obtained subsequently.

(March 2008). Discoveries so far made are described in Chapter 11, but one example is given in Figure 9.2.

The dots in Figure 9.2 show the first observational data obtained by transit photometry. The planet of HD209458 had previously been detected by Doppler spectroscopy (Section 10.2). The transit data were obtained in 1999 by means of the 100 mm aperture telescope of the STARE project at Boulder, Colorado. Such a telescope can be regarded as a medium sized camera. The HST subsequently observed HD209458 and obtained the labelled light curve. The scatter of the HST measurements is about the thickness of the line – rather less than the scatter in the STARE data!

Most of the transit searches in progress are from ground level. Like STARE, not all of these involve large telescopes. Of particular note is SuperWASP (WASP = Wide Angle Search for exoPlanets). This is operated by a consortium of eight academic institutions, seven in the UK and one in Spain. SuperWASP consists of two instruments each of which is operated remotely. One is located high on the island of La Palma in the Canaries, the other is located at the South African Observatory.

Figure 9.3 shows the one in South Africa. You can see that it consists of eight telescopes. In fact these are standard cameras, each with an aperture $D = 200$ mm. Such a small aperture puts faint stars beyond the reach of SuperWASP, but the eight cameras between them cover a huge field of view – roughly 2,500 times the area of the Moon in the sky. The SuperWASPs can thus survey much of the sky, and they are doing so, repeatedly. Over the few years it has been operating, brightness variations in many stars have been recorded, but as of October 2007 just three of these have proved to be planets, giant planets, though there are a few promising candidates awaiting confirmation of planetary mass by Doppler spectroscopy. Many more planets are expected to be discovered in the years to come.

FIGURE 9.3 SuperWASP in South Africa. The remotely operated instrument consists of eight cameras, each with an aperture of 200 mm, each looking in a different direction. (David Anderson, Keele University)

Space missions are likely to discover many planets through transit photometry. One of these, COROT, led by the French, launched in December 2006. Even though its aperture is only 270 mm, the advantage of being above the Earth's atmosphere is manifest in its ability to detect brightness variations of only a few parts per million for bright stars, particularly if many transits of a star are observed and the results added together. It will thus be able to detect planets with radii as small as that of the Earth out to thousands of light years. It also has a large field of view, about 50 times the area of the Moon in the sky. In May 2007 it made its first discovery, a planet with a radius 1.5–1.8 times that of Jupiter, the uncertainty lying in the radius of the star. It orbits its G dwarf, which is about 1,500 light years from us, in only 1.5 days, so the planet must be very hot. Ground-based Doppler spectroscopy (Section 10.2) has shown the planet's mass to be 1.3 times that of Jupiter. Many more discoveries are expected by the end of its nominal two year mission.

Another notable space mission that will detect planets in transit is a NASA Discovery mission called Kepler, due to launch in November 2008. It will have a 950 mm aperture, a field of view of about 400 times the area of the Moon in the sky, and very high photometric precision. Kepler will push transit photometry towards its limits, and in surveying about a hundred thousand stars could discover tens of thousands of planets, including Earth-twins and even Mars size planets closer to the star.

Both COROT and Kepler could detect exoplanets around F, G, and K dwarfs out to many thousands of light years.

9.2   GRAVITATIONAL MICROLENSING

This method relies on the effect of an exoplanetary system on the light we receive from a star beyond it, called the background star. It involves concepts that you might find difficult, in which case it will suffice for future chapters if you just get the gist of the method.

Figure 9.4 shows an exact alignment between a background star, an interposed star that might have a planet, and us as observers. You might think that the background star would be hidden from view. Instead, the gravity of the interposed star bends the light that passes near it. Some of the bent light reaches us in the manner shown, and we would see the background star, if we had sufficient resolution, as a ring of light around the interposed star. The interposed star has thus acted as a gravitational lens, and is called the lensing star. The ring is called the Einstein ring, after Albert Einstein, whose theory of general relativity (1916) explains the bending in detail.

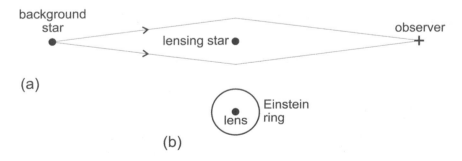

FIGURE 9.4 (a) Gravitational lensing with exact alignment of the observer, the lensing star, and the background star. (b) The image of the background star – an Einstein ring.

Box 9.1 Einstein and General Relativity

Albert Einstein (1879–1955) was a German–Swiss–American theoretical physicist who ranks with Isaac Newton in his impact on our understanding of the nature of the physical world.

Among his many profound achievements was the theory of general relativity that he published in 1916. It is essentially a theory of gravity. Previously, for nearly 300 years Newton's theory of gravity seemed able to account for almost all observations involving gravity, such as the flight of projectiles on Earth, and the motion of planets and other bodies such as comets, around the Sun. Note the "almost all". There were some small discrepancies that do not occur in Einstein's theory, such as fine detail in the orbit of Mercury.

Both theories predict that mass bends light, but Einstein's theory predicts an effect twice the size of that predicted by Newton's theory. Measurements of stars in directions very close to the edge of the Sun show that it is Einstein's theory that gives the observed displacement of the stars from their true positions.

Albert Einstein in 1921.

The Einstein ring has a radius that increases as the mass of the lensing star increases and depends also on the distances from us to the lensing star and to the background star. For typical distances and masses, the *angular* radius of the Einstein ring as we observe it, is smaller than the angular radius of the central fuzzy disc of a stellar image produced by a telescope with a main mirror with a diameter even as large as 10 meters. Therefore it cannot be directly imaged. The lensing can however be detected because we receive more light from the background star when it is lensed than when it is not lensed. Thus, as the stars' motions across the sky produce the alignment we would observe an apparent brightening of the background star. Detection of gravitational lensing via apparent brightening is called gravitational microlensing.

An analogy might help. Get hold of a magnifying glass and view a distant small source of light through it (*not* the Sun, which would damage your eyes). Vary the distance between the glass lens and your eye, and there will be a point where the lens seems full of light, much more light than when the source is viewed without the lens. The lens has directed light into your eye that otherwise would have missed it. Like all analogies this one has shortcomings. The main one is that the distance between the glass lens and your eye cannot be varied much without losing the enhancement. With a gravitational lens this distance is not critical.

Alignments of the background star and the lensing star are never exact. Figure 9.5(a) shows a sequence where the actual positions of the background star (with respect to the lensing star) are the five small circles. The lensing produces a

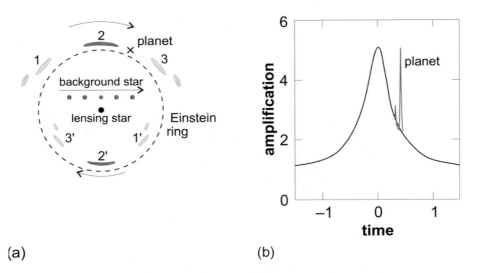

(a) (b)

FIGURE 9.5 (a) Gravitational lensing with inexact alignment, where corresponding pairs of background star images are shown e.g., at (1, 1′), (2, 2′), (3, 3′). (b) The light curve from the inexact alignment, where the red lines are the effect of a planet at 'x'.

sequence of pairs of distorted images e.g., at (1, 1′), (2, 2′), (3, 3′). Note that these are not far from where the Einstein ring would have been. The light received from these images of the background star is shown by the black line in Figure 9.5(b). Remember that we cannot see the detail in the images – we just see a dot of light that grows in brightness until the closest angular approach of the background star to the lens, and then the brightness declines. This variation is called the amplification – it equals 1 when there is no lensing.

The amplification depends only on the angular separation of the stars as we see it, expressed as a fraction of the angular radius of the Einstein ring – it does not depend on the mass of the lensing star. The encounter lasts for a duration depending on the motion of the lens across the sky with respect to the source. For typical values this is a few days.

Microlensing by a planet

A planet can also produce lensing of the light from a background star. Moreover, because the peak amplification depends only on the closest angular approach expressed as a fraction of the angular radius of the Einstein ring, a planet could produce as large an amplification as a star. Unfortunately, because the radius of the Einstein ring decreases as the mass of the lens decreases, the target area is small for a planet, and the likelihood of a sufficiently good alignment is correspondingly small. The duration also decreases as the mass of the lens decreases, and so even if there is excellent alignment, the lensing is over in a matter of hours or minutes, and is therefore easily missed.

The practical approach is therefore to detect the beginning of lensing by a star, and use this as an alert to look for modifications produced by any planet orbiting the star. This will only be appreciable if the planet is located near the path of the images of the source star, for example at 'x' in Figure 9.5(a). In this case the planet will deflect the light from the image of the background star, which will modify the light curve as shown by the red lines in Figure 9.5(b). These form narrow peaks. Their height is greater the closer the planet is to the image trajectory, and is independent of the mass of the planet.

Given that the planet has to be located near the path of one of the source images, and that these are near the Einstein ring, it is clear that the separation of a detectable planet from its star has to be such that, when projected on the plane of the sky, it lies near the Einstein ring. Therefore, only a small proportion of exoplanets will be discovered by gravitational microlensing.

Nevertheless, microlensing is an important technique, for several reasons. First, it can find planets up to tens of thousands of light years away. This is far further than can be reached by transit photometry and by the techniques to be described in Chapter 10. It can also find planets around intrinsically less luminous stars than is the case for other techniques.

Second, if, as well as the duration of the stellar and planetary events, we know the mass of the lensing star, it turns out that we can get the mass of the planet. There are several ways in which observations of a star can give us its mass.

Moreover, the duration of the microlensing decreases rather slowly as mass decreases. For example, though the Earth is 318 times less massive than Jupiter, the duration of microlensing by the Earth is only 18 times less than that of Jupiter. This, plus the fact that the height of the planet's peaks is independent of its mass, means that the prospects for detecting Earth mass planets are enhanced.

Third, we do not have to make observations over one or more orbits of the planet – a planet in a 5 AU Jupiter orbit, which has an orbital period of 11 years around a solar mass star, could be detected in days rather than years. Finally, satellites of a planet could be detected via subsidiary peaks.

There are, however, limitations. First, we learn little about the planet's orbit. We know that the angular separation on the sky of the star-planet distance during the microlensing is roughly the angular radius of the Einstein ring. If we know the distances to the stars, which we usually do well enough, then we can get the linear radius of the ring. This is typically a few AU. So all we learn is that the planet at some instant was at a projected distance of a few AU from its star – we have no way of obtaining the actual distance, in three dimensional space, between star and planet. Second, lensing is one-off – follow up lensing events are extremely unlikely, though if the system is near enough it could be explored by some of the other methods of discovery and investigation.

Search programs

A number of microlensing surveys have taken place, in directions in space in which there are many background stars to act as sources and many closer stars to act as lenses, such as towards the nuclear bulge of the Galaxy (Section 7.1). You will appreciate that lensing surveys will also pick up transits, and this has occurred on many occasions.

Very many lensing events have been recorded, for example, in the OGLE (Optical Gravitational Lensing Experiment) survey that for some years has been using the 1.3 meter Warsaw telescope at Las Campanas Observatory in Chile. This has also observed planets in transit.

More recently PLANET has been coordinating a network of five telescopes ranging from 1–2 meters in aperture, distributed in longitude around the southern hemisphere in order to perform almost continuous monitoring in the direction of our Galaxy's nuclear bulge. Among other discoveries it has found a planet only about five times the mass of the Earth. In 2005 it joined forces with RoboNet-1.0, a network of three UK operated robototic telescopes with 2.0 meter apertures.

In gravitational microlensing early alerts are particularly important so that many observations of any subsequent planetary event can be made. The systems for implementing such alerts are now in place.

The handful of planets so far discovered by gravitational microlensing will be included in Chapter 11.

10

Searching for exoplanets by detecting the motion of their stars

In the previous chapter I concentrated on those indirect methods where we infer the presence of a planet from its influence on the quantity of radiation we receive from this star or from some background star.

In this chapter I describe the two indirect methods that depend on the influence of an exoplanet on the motion of the star it orbits. The second of these methods (Doppler spectroscopy) has discovered most of the exoplanets so far known.

10.1 ASTROMETRY

Stars move steadily through space, but though their speeds can be large by terrestrial standards, they are at such huge distances that with the unaided eye we do not perceive the motion of one with respect to another, not even in the course of a human lifetime. Some such motion would be apparent for some stars if we lived considerably longer, but even the peoples of the ancient world would recognize the stellar constellations as we do, even though in many cases a discernable change in shape has since occurred. But telescopes that magnify star fields and use sensitive position measuring techniques show that all of the stars are moving with respect to each other.

The record for motion across the sky is currently held by Barnard's Star, which is 5.94 light years away, making it the fourth nearest star to the Sun, after the triple star system Proxima Centauri-Alpha Centauri A-B (Figure 7.4). Its closeness contributes to its rapid motion across the sky. It moves such that it changes its direction with respect to the cosmic background at a rate that would traverse the Moon's diameter in the sky, 0.5°, in about 170 years. The cosmic background is a reference frame for directions in the sky, centered on the Earth. For practical purposes it can be regarded as the framework provided by distant galaxies.

Barnard's Star is much too faint to be seen with the unaided eye. Alpha Centauri, though faint, can be seen under dark sky conditions, as a single point (the two stars, A and B, are too close together to be seen separately). It moves across the sky nearly three times slower than Barnard's star but this would be easily noticeable in a thousand years, during which time it would move through an angle of just over 1° with respect to the cosmic background.

Box 10.1 The measurement of angles

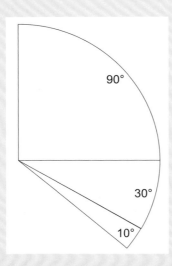

It is very common in astronomy to express distances in terms of angles. For example, if you look to the horizon and then tilt your head so that you are looking upwards, you will have moved your gaze through ninety degrees of arc across the sky, written 90°. If you hold out your hand at arm's length, palm upwards, with the fingers bent toward you, the angle across the base of your fingers is about 7°. When you look up into the sky at a full Moon the angle across the Moon's diameter will be about 0.5°. This is the Moon's angular diameter. Lunar craters have yet smaller angular sizes. Many angles in astronomy are extremely small, so the degree of arc is subdivided into 60 minutes of arc (arcmin), and an arcmin is divided into 60 seconds of arc (arcsec).

Various angles, measured in degrees of arc (°).

Note that an angle is determined by the actual size of an object and by its distance. The angular diameter of the Sun is about the same as that of the Moon, even though the Sun's actual diameter is nearly 400 times that of the Moon. This is because the Sun is also nearly 400 times further away. This is a coincidence, and was not true in the distant past when the Moon was closer to the Earth.

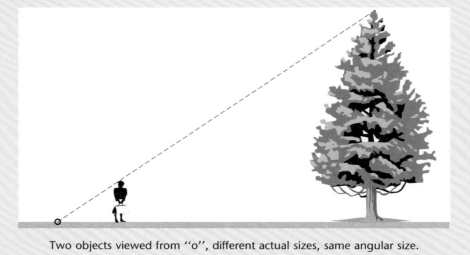

Two objects viewed from "o", different actual sizes, same angular size.

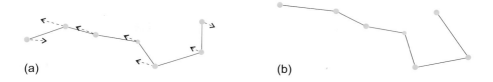

FIGURE 10.1 The Plough (The Big Dipper) (a) today (the arrows show the present motion across the sky) (b) 52,000 years ago.

The great majority of stars visible to the unaided eye move far more slowly than Alpha Centauri. Figure 10.1 shows the brightest stars in the well known pattern of stars called The Plough (also known as the Big Dipper). This is part of a much larger constellation called Ursa Major (Latin for The Great Bear). You can see that even 52,000 years ago the shape was not beyond recognition.

The angular motion of a star across the sky with respect to the cosmic background is called its proper motion, and it is measured in seconds of arc (arcsec) per year. The stars also move along our line of observation. This is called radial motion, and is usually measured in km per second. But first, let's see how we use a star's proper motion to detect whether it has planets, in a technique called astrometry.

Detection of planets by astrometry

"Stars move steadily through space" – almost true, but not quite! If a star has a companion, be it another star in a binary system or a planet, then both bodies orbit what is called the center of mass of the system. Figure 10.2 shows two stars orbiting each other where one star is twice the mass of the other. In this case the more massive star is half as far from the center of mass than the less massive star – it is the less massive star that moves more. A simple model of this system can be constructed from two balls and a thin rod, the one ball being twice the mass of the other, for example 0.2 kg and 0.4 kg. If you built such a model you would find that the balance point is twice as far from the 0.2 kg ball than from the 0.4 kg ball. This balance point is also the center of mass of the two balls.

When the binary system moves through space it is the center of mass that moves steadily. Figure 10.3 shows the small circular orbit of a star due to a planet a tenth of its mass, in a circular orbit. The star's proper motion is a combination of its orbital motion and the motion of its center of mass. By measuring the position of the star often enough to reveal its complicated path, we could extract the orbital component of its motion and hence deduce the presence of the planet even if it were not visible.

Unlike the example in Figure 10.2, the motion of a star induced even by a planet of Jupiter's mass, will be very slight. The position of the center of mass is given by

mass of body A × distance of A = mass of body B × distance of B

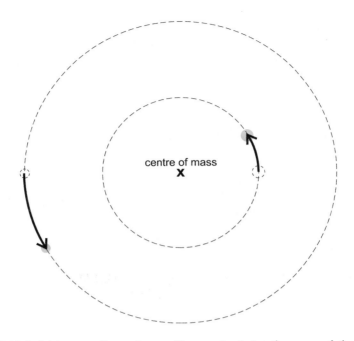

FIGURE 10.2 A binary stellar system, with one star twice the mass of the other. The center of mass is the point around which the two stars move in their orbits (in this case circular).

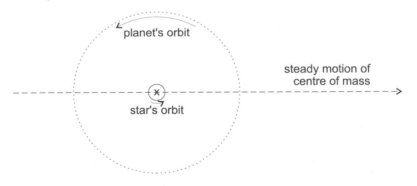

FIGURE 10.3 A star with a planet a tenth of its mass in a circular orbit. The center of mass is moving through space at a steady rate.

where "distance" refers to the distance from the center of mass. You can test this formula by making the two masses equal. The two distances are then equal i.e., the center of mass is half way between body A and body B, as you would expect. You can also check that if one mass is twice the other then the center of mass is twice as close to the greater mass. The Sun is 1047.6 times the mass of Jupiter, the most massive planet in the Solar System. Our formula now tells us that, if Jupiter were the only planet in the Solar System, it would orbit 1047.6 times further from the center of mass of the Solar System than the Sun does. The presence of the other planets complicates the Sun's orbit, mainly because of Saturn, which, at a little under a third the mass of Jupiter, is the second most massive planet in the Solar System.

The dashed line in Figure 10.4 shows the Sun's orbital motion that would be observed face-on from 30 light years away were Jupiter the only planet in the Solar System. Note how small the Sun's angular excursion is – it moves to each side of its mean position by just a little over 0.5 *milli*arcsec (0.0005 arcsec). Compare this with the angular diameter of the Moon, which is 0.5°, or 1,800 arcsec. Clearly it is not easy to measure the motion of a star caused by the planets in orbit around it. The actual motion of the Sun, as would be observed from 30 light years with the effects of all the planets included, is also shown. The overall excursion is much more complicated, but only a little greater than that with Jupiter alone.

If a stellar orbit is not presented face-on to us the question arises of whether we can deduce the face-on view. The answer is that we can. This is achieved

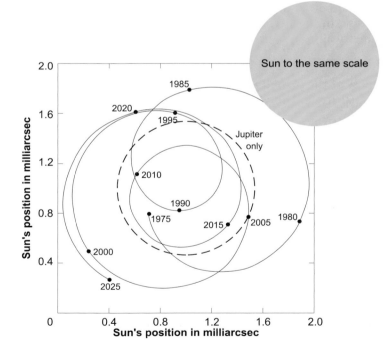

FIGURE 10.4 The angular size of the Sun's orbital motion as seen face-on from a distance of 30 light years (solid line) and were Jupiter the only planet (dashed line). The angular size of the Sun viewed from 30 light years is also shown – the center of mass of the Solar System does not stray far from the Sun's surface.

angular size   actual size

point of
observation

distance

actual size = angular size × distance

FIGURE 10.5 Obtaining the actual size of an orbit from its angular size. (The formula requires that the angular size is measured in radians, where 1 radian = 57.3°.)

through observations of the rate at which the star traverses its orbit, though the details will not concern us.

What does concern us is what we learn about a planet from astrometry. Astrometry gives us the size of the star's orbit in angular units. We need to know the distance to the system so that we can obtain the actual size of the star's orbit (e.g., in AU), as illustrated by Figure 10.5. If we also know

- the mass of the *star* (obtained from observations of its radiation – details beyond our scope); and
- the orbital period of the *planet* (from Figure 10.2 you can see that this is the same as the orbital period of the star, which has been determined astrometrically)

then it turns out that we can calculate

- the mass of the planet; and
- the actual size of the planet's orbit.

I'll not go into the algebraic details. The eccentricity of the star's orbit is the eccentricity of the planet's orbit. When there is more than one planet the star's orbit is more complicated (Figure 10.4).

You can probably see that the detection of a planet is easier

- the more massive it is compared to its star (thus making the star move in a larger orbit);
- the larger its orbit (examine the word equation on page 137); and
- the closer the system is to us (for a given orbit, its angular size will be larger).

What has been discovered so far?

Results from astrometry

Attempts to detect stellar orbits due to planets have a long history. The first major attempt was by the Dutch astronomer Peter van de Kamp. He began his search when he joined the Sproule Observatory in Pennsylvania in 1937. He used a refracting telescope i.e., one using lenses rather than mirrors, with a main lens aperture D of 0.61 meters. At visible wavelengths the angular radius of the central fuzzy disc in the diffraction limited case (Figure 8.2(b)) is then around 0.2 arcsec.

By estimating the position of the center of the fuzzy disc the ultimate positional resolution could, in principle, be a few times better than this. But you can see From Figure 10.4 that this would be nowhere near good enough to detect the Sun's motion from 30 light years. What about easier targets, such as Barnard's star, only 5.94 light years away?

He recorded the positions of stars on photographic plates over several decades. He became convinced that he had detected planets around a few stars, including Barnard's Star. A major overhaul of the telescope in 1949 caused him to exclude from analysis the plates he had obtained before that date, but further telescope adjustments in 1957 caused more uncertainties. His positive results were challenged by others, partly on the basis of the 1957 adjustments, but partly because of problems caused by his underexposure of photographic plates and his use of rather few reference stars to provide a (nearly) fixed distant background. Nevertheless, at his death in 1995 he still believed that Barnard's Star has at least one planet. This claim has not been confirmed by others, though the data do show some indication of oscillatory motion of the star.

Others have attempted astrometry from the ground using single aperture techniques like van de Kamp, more recently replacing the photographic plate with a CCD. No definite discoveries have been made, though a previously known giant planet, orbiting the M dwarf Gliese 876, was detected by astrometry in 2002 by the HST (Hubble Space Telescope). This detection was aided by the low mass of the star, about a third the mass of the Sun, and its consequently greater orbit, and by its proximity, just 15.4 light years away.

Ongoing single aperture searches include the Carnegie Astrometric Planet Search, which uses a 2.5 meter aperture telescope at Las Campanas in Chile, and the STEPS (Single Telescope Extrasolar Planet Survey) project using the 5.1 meter (200 inch) Hale telescope on Mount Palomar in California. STEPS has a candidate planet, which would make it the first discovery of an exoplanet by astrometry, but further observations are needed to confirm the object as a planet rather than a low mass star.

The Earth's atmosphere poses two big problems for ground-based astrometry. First, atmospheric refraction depends on the altitude of an object above the horizon, and therefore introduces distortion across an image that varies with the altitude at image center. Second, atmospheric winds and turbulence (Section 8.2) cause a jitter in the positions of star images that limits the precision of measurement of star separations to about 0.1 arcsec even at high altitude observatories where the effect of the atmosphere is reduced.

There is a reduction in jitter if large apertures are used. For a 10 meter aperture telescope on a high altitude site, angular precisions of 100 millionths of an arcsec, or better, can be obtained with exposure times of the order of an hour. Jitter is further reduced if adaptive optics is used. Note that this is good enough to detect the effect of a Jupiter-twin at a distance of 30 light years from us. (Recall that a Jupiter-twin is a Jupiter *mass* planet about 5 AU from a solar type star i.e., a G dwarf, about the same age as the Sun.)

Forthcoming astrometry on the ground and in space

The two Keck telescopes on Mauna Kea and the four VLT telescopes in Chile are 8–10 meter aperture telescopes at high altitudes, and are capable of operating interferometrically. The greater positional resolution of interferometers will give them the capability to detect planets astrometrically.

At the VLT, astrometry will be being performed with the four telescopes, each using adaptive optics, in a system called PRIMA (Phase Referenced Imaging and Microarcsecond Astrometry), due in 2008. PRIMA will have an astrometric precision of 10 millionths of an arcsec. The semimajor axis of the star's orbit would have to be about twice this for the orbit to be determined with useful accuracy. PRIMA will have the capability to detect the stellar motion caused by a Jupiter-twin out to about 800 light years. Earth-twins will not be detectable (Earth mass planets, about 1 AU from solar-type stars of the same age as the Sun), but Earth mass planets several AU from nearby low mass M dwarfs would be. Unfortunately, several AU is well beyond the habitable zone of an M dwarf.

Developments are needed at the Keck telescopes before substantial surveys for planets can be made.

Further in the future, around 2010, an array of radiotelescopes operating at millimeter wavelengths will have the interferometric capability to discover planets astrometrically. This array, ALMA, will consist of 64 antennas and be located in Chile. It will have an astrometric precision of about 100 millionths of an arcsec, which will give it the capability of detecting the effect of a Jupiter-twin out to many tens of light years.

Space has the huge advantage for astrometry that the deleterious effects of the Earth's atmosphere are absent. The potential of space-based astrometry was demonstrated by ESA's (European Space Agency) Hipparcos spacecraft, which was launched in 1989 on a four year mission. Its best precision was about 500 millionths of an arcsec, on the threshold of being able to detect a Jupiter-twin at a distance of 30 light years.

In the near future several space missions should be capable of the astrometric detection of planets. These include the NASA mission SIM (Space Interferometry Mission), under study for launch at some unknown time after 2009, and the ESA mission Gaia, scheduled for launch in 2011. Both have two telescopes. SIM will be a targeted mission, and will be able to achieve precisions of a few millionths of an arcsec even for faint stars. These faint stars would require an exposure time of several hours per single observation, and a few tens of observations per star to get the orbit. Jupiter-twins could be detected out to a few thousand light years, and Earth-twins to a few tens of light years, but relatively few stars could be targeted per year.

Gaia is a sky survey mission, rather than a targeted mission. Figure 10.6(a) shows how it will look in orbit, and Figure 10.6(b) shows its optical heart. There are two astrometric telescopes, each with mirror apertures 1.45 by 0.5 meters. These will enable very accurate separations between widely separated stars to be obtained. Over the nominal five year mission, the same fields of view will be

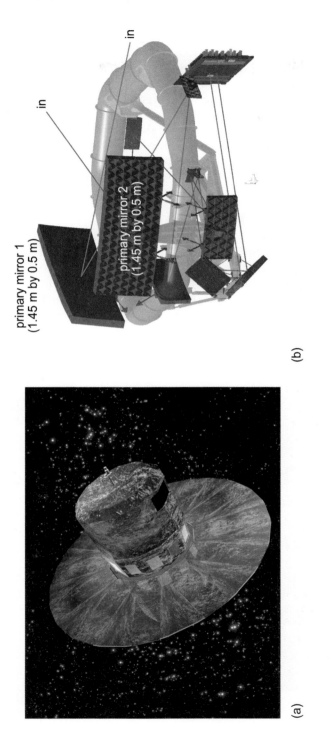

(b)

(a)

FIGURE 10.6 (a) Gaia as it will appear in orbit. (ESA C Carreau) (b) The optical heart of Gaia, showing the two primary mirrors 1.45 m by 0.5 m used for astrometry. (EADS Astrium)

recorded about 80 times. For bright stars (brighter than the Sun would be at a distance of 350 light years), angular excursions of the star as small as 40 millionths of an arcsec could be measured with reasonable accuracy. This is the angular width of the finest human hair, 0.02 mm, at a distance of 100 km – pretty well unimaginable. It will be sufficient for a Jupiter-twin to be detected out to about 350 light years, and to detect planets several times the mass of the Earth at 1 AU from M dwarfs out to a few tens of light years. Earth mass planets will be beyond detection.

A third telescope on Gaia will be used to measure stellar velocities along the line of sight. Though Gaia will not attain sufficient precision to detect planets this way, this is the basis of the Doppler spectroscopy technique, to which I now turn.

10.2   DOPPLER SPECTROSCOPY

Doppler spectroscopy is the technique that has been used to discover nearly all of the known exoplanetary systems. It relies on the radial velocity of a star – its motion along our line of sight. Unlike its proper motion, this component of a star's motion causes no change in its position on the sky. How then, can we measure its radial velocity, and, if it has one or more planets, the variation in the radial velocity as the star moves around the center of mass of the system?

The Doppler effect

The answer lies in the Doppler effect, named after the Austrian physicist Christian Johann Doppler. It is a phenomenon associated with waves. It applies

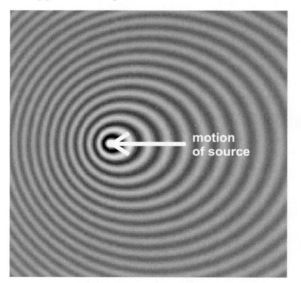

FIGURE 10.7 Waves from a moving source.

to all types of wave, including electromagnetic waves. If the source of the waves is moving toward the observer the waves arrive closer together than they would if there were no such motion. If the source is moving away from the observer the waves arrive further apart than they would if there were no such motion. In other words, the rate of arrival of the waves depends on the motion of the source with respect to the observer in the radial direction to the observer. This is the Doppler effect. Figure 10.7 illustrates it for a single wavelength.

You will surely have experienced this phenomenon in the rise and fall of the pitch of a siren on an emergency vehicle as it sweeps past. To an earlier generation, the scream of a train whistle as the train rushed past also displayed this phenomenon. The waves in these two examples are sound waves, the rate of arrival of which we perceive as pitch.

Box 10.2 Doppler and the Doppler effect

The Austrian physicist Christian Johann Doppler (1803–1853) predicted the effect named after him, in 1842. In 1845 the Doppler effect was experimentally verified in Holland, by the use of a locomotive drawing an open car with several trumpeters.

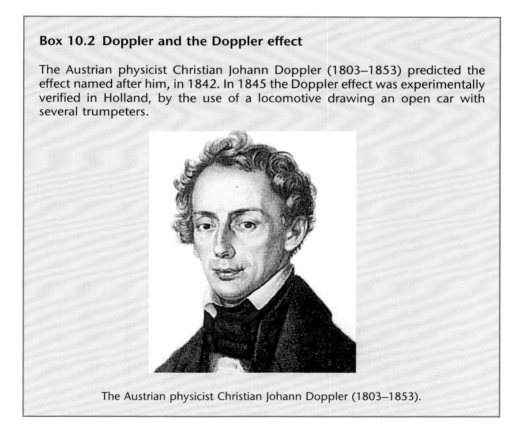

The Austrian physicist Christian Johann Doppler (1803–1853).

Detecting planets by Doppler spectroscopy

In the case of starlight, the question arises, what is the analog of a note with a pitch? The answer is, one of a star's spectral absorption lines. The electromagnetic radiation emitted by a star covers a wide range of wavelengths, as you saw for the Sun in Figure 2.10. These are emitted from the photosphere of the star with no sharp changes with wavelength. As the radiation traverses the star's atmosphere, ions (and some atoms) in this atmosphere absorb the radiation at certain wavelengths, each one characteristic of the chemical element to which the atom or ion belongs. This results in the star looking dimmer at certain wavelengths than at adjacent wavelengths. The deficit is called an absorption line. This is illustrated for a simple case in Figure 10.8. Displayed in this way you can see that the absorption is more of a narrow dip than a line.

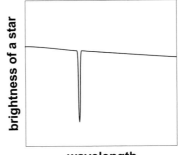

FIGURE 10.8 A single absorption line in a narrow range of wavelengths.

Figure 10.9 is for real data. Note how many absorption lines there are even in this small span of wavelengths. This is a consequence of the many different ions (and a few atoms) in the star's atmosphere, each with many absorption lines. It is the narrowness of the lines that explains why they are not apparent in the rather coarse solar spectrum in Figure 2.10 – at finer detail the Sun's many absorption lines become apparent.

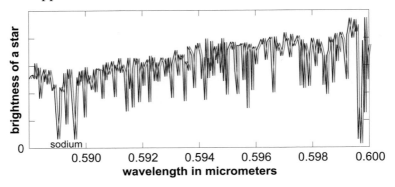

FIGURE 10.9 Absorption lines in part of the spectrum of a star, over a narrow range of visible wavelengths. The red line shows the spectrum Doppler shifted to longer wavelengths because the star is moving away from us. It is these shifted lines that we observe.

The spectrum shown by the black line in Figure 10.9 is for zero radial velocity of the star. The wavelengths of the absorption lines in this case are the same as those obtained in the laboratory. The spectrum shown by the red line is for the star moving *away* from us. This corresponds to the *right* half of Figure 10.7 – the wavelength of each spectral line is increased. If the star is moving *toward* us, then each electromagnetic wavelength emitted by the star displays the phenomenon illustrated in the *left* half of Figure 10.7 – the wavelength is reduced.

Now suppose that the speed of motion in the radial direction varies. This will cause the increase (or decrease) in wavelength to vary. By measuring the wavelengths repeatedly, it is possible to discern the variation in radial velocity. Of particular interest to us is the orbital motion of a star caused by the orbital motion of its planets. Star and planets move around the center of mass of the system. This will be superimposed on the steady motion of the center of mass in the radial direction.

Figure 10.10 shows the simple case resulting from a star orbiting in a circle around the center of mass of a system that has just one planet in a circular orbit. The left hand axis is labelled with the change in wavelength of a spectral line – the observed wavelength minus the wavelength at the source in the stellar atmosphere (the steady motion in the radial direction has been extracted). The right hand axis shows the associated cyclic variation in the radial velocity, which is readily calculated from the cyclic variation in wavelength. The graph also shows the period of the variation, which is the period of the orbits of the star and planet around the center of mass.

From the range of variation of the radial velocity, and the period of the variation, we can obtain

- the mass of the planet; and
- the semimajor axis of the planet's orbit.

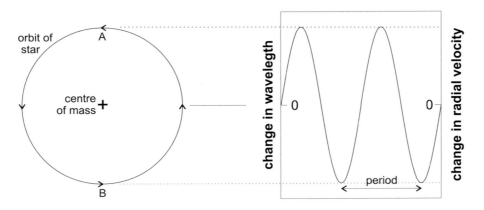

FIGURE 10.10 A star in a circular orbit presented edgewise to the observer, and the associated changes in spectral line wavelength and radial velocity.

As with astrometry, we need to know the mass of the star to get the planet's mass. We also need to know the star's mass to calculate the semimajor axis of the planet's orbit from the period. As mentioned earlier, the mass of a star can be obtained from various analyses of its radiation.

The shape of the graph in Figure 10.10, which is called a sinusoid, is a direct consequence of the orbit being circular. Figure 10.11 shows some real data for three stars each with a single giant planet. The measurements have uncertainties, and therefore they do not all lie on smooth curves. You can, however, see that only the top curve is sinusoidal, indicating a very low eccentricity. The other two curves are distinctly not sinusoidal. The shape depends on the eccentricity of the star's orbit. This is the same as that of the planet's orbit. Thus we can also obtain

- the eccentricity of the planet's orbit.

The caption to Figure 10.11 gives the eccentricities. Table 10.1 gives further data on these systems.

Clearly, the greater the variation in the star's radial velocity the easier it will be to detect it, and the more accurately it can be measured. This variation increases as

- the ratio of the mass of the planet to the mass of the star increases
- the size of the planetary orbit decreases – the orbital period is then less and so the speed of the star (and planet) around its orbit increases.

A small period also facilitates detection because cycles of data are then acquired more rapidly.

TABLE 10.1 Properties of the three systems in Figure 10.11

Star properties star	Mass (solar masses)	Planet properties Distance (light years)	Mass (Jupiter masses)[1]	Semimajor axis (AU)	Eccentricity
51 Pegasi	1.11	48	0.468	0.052	0.0197
70 Virginis	1.1	72	7.44	0.48	0.4
16 Cygni B	1.01	69.8	1.68	1.68	0.689

(1) This is the minimum mass of the planet.

The orbital inclination problem

The case shown in Figure 10.10, and the associated discussion, is for the orbit presented edge-on to the observer. What if the orbit were presented face-on? In this case there is no orbital motion toward the observer, and no matter how rapidly the star dashed around its orbit we would be unable to detect its motion by Doppler spectroscopy. Any planet would be undetected.

Figure 10.12 illustrates the general case, where the star's orbit is presented to us neither edge on nor face-on, but at some in between angle. This angle is called

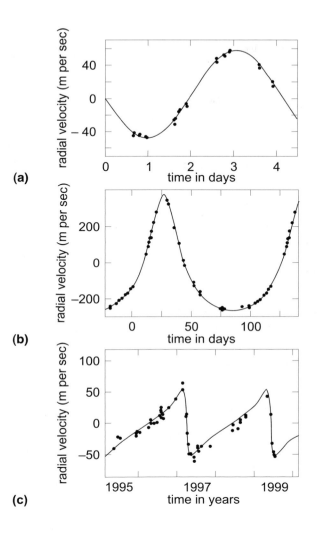

(a)

(b)

(c)

FIGURE 10.11 Measurements of the radial velocities of three stars (a) 51 Pegasi, near-circular orbit, eccentricity 0.0197 (b) 70 Virginis, orbital eccentricity 0.4 (c) 16 Cygni B, orbital eccentricity 0.689, differently oriented toward us than 70 Virginis. The dots are the measurements; the curves are theoretical fits. The radial velocities of the center of mass have been subtracted.

the inclination of the *star's* orbit with respect to the plane of the sky. Its extreme values are 0° for a face-on presentation, and 90° for an edge-on presentation. With the planet's orbit in the same plane as the star's orbit this angle will also be the inclination of the planet's orbit with respect to the plane. We need to know this angle so that we can obtain the radial velocity variations that we would have

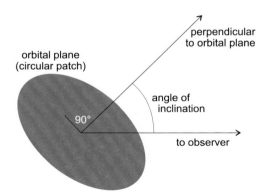

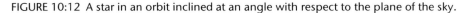

FIGURE 10:12 A star in an orbit inclined at an angle with respect to the plane of the sky.

observed in an edge on presentation – we are only measuring the component of the variation that is directed toward us.

There are a few ways to estimate the inclination of the star's orbit. If the star has a visible circumstellar disc of gas or dust, then, given that planets were formed earlier from such a disc, they are likely to be orbiting in the same plane as the disc. If images of the disc show it as elliptical, then, given it is likely to be circular, we must be seeing it at an angle. This angle can be estimated and it is likely it will be the inclination of the star's orbit and therefore of the planets' orbits. The inclination can also be estimated in some cases by observations of the star that tell us the inclination of its rotation axis with respect to the plane of the sky. Assuming that the planets orbit in the equatorial plane of the star, we get the inclination. This assumption, however, might be poor. The most certain way of obtaining the inclination is when a planet is observed in transit. This shows that its orbital inclination with respect to the plane of the sky is close to 90° (edge on presentation).

If we cannot measure or estimate the orbital inclination, then, from the measured radial velocity variations, we get the mass of the planet were it to be in an edge on orbit. Its actual mass can only be greater, corresponding to its full orbital velocity, and not just the component directed toward us. Thus, for a given *measured* radial velocity variation of the star, the *actual* orbital velocity will be larger, requiring a more massive planet to swing the star around faster. Doppler spectroscopy alone thus gives us the minimum mass of the planet. This is the case for the three systems in Table 10.1. The implications of this restriction for population statistics of actual masses are presented in Section 11.2.

Obtaining stellar spectra

How are stellar spectra of the sort in Figure 10.9 obtained?

In order to obtain a star's spectrum, the light collected from the star by a telescope has to be passed through a device that sends different wavelengths into

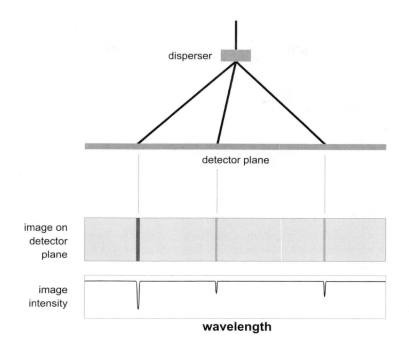

FIGURE 10.13 A dispersing device separating three absorption lines, with (below) the image intensity obtained by scanning across these lines.

different directions so that they fall in different positions on a detector. This dispersion, as it's called, is illustrated in Figure 10.13 in the simple case when a source has just three well-separated absorption lines. A familiar disperser is a glass prism (Figure 10.14), though it does not spread apart wavelengths widely enough to be useful. A better device is called a diffraction grating, which consists of many parallel rulings with spacings not very different from the wavelength range of interest in the spectrum. This spreads different wavelengths much further apart enabling absorption lines of almost the same wavelength to be seen separately. Any device which displays a spectrum is called a spectrograph.

Once the stellar spectrum is produced we have to repeatedly measure the wavelength of at least one spectral line to detect any variations in radial velocity. In practise very many lines are measured to increase accuracy. The stellar wavelengths are measured by comparing their positions on the detector with those from a reference source in the observatory that has numerous narrow, accurately known and stable spectral lines over a wavelength range where the star also has many lines. A commonly used source is a gas of molecular iodine. This has many extremely narrow spectral lines over the range 0.5–0.6 micrometers that are very stable in wavelength.

From the varying differences in wavelength between the iodine and the star, the variations in radial velocity are derived. But there is one further step – we

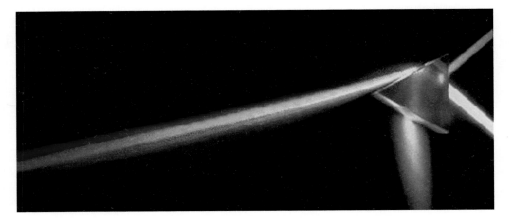

FIGURE 10.14 A glass prism, sending different visible wavelengths into different directions to the left from a narrow beam of white light coming in from the right. (The white light emerging at the bottom and top of the prism is from reflections.) (NASA)

have to subtract the Earth's varying radial velocity with respect to the star, arising from the Earth's orbital motion. When this is done we have the radial motion of the star with respect to the center of mass of the Solar System, which moves through space at a sufficiently constant speed to obviate the need for further adjustments.

Radial velocities are currently being measured with a precision approaching 2 meters per second. This is certainly enough to detect a Jupiter-twin, provided that the observations are made for at least a large fraction of its 11.9 year orbital period. It will be difficult to push this precision to the 0.1 meters per second needed to detect Earth-twins, though an Earth mass planet in the habitable zone of a low mass M dwarf could produce a variation of several meters per second, and so would be detectable. This is because of the low mass of the star and the proximity of the habitable zone to it. The former increases the ratio of the mass of the planet to the mass of the star, and the latter corresponds to short orbital periods.

The ultimate limit might be set by stellar activity. For example, the convective cells in a star's outer regions also cause Doppler shifts, and the convective activity will vary with the stellar cycle, thus giving the appearance of a varying radial velocity. For a solar type star the variation will be around one meter per second, and this might stand in the way of unambiguous interpretation. For many M dwarfs, notably when they are young, the variation is probably higher, which would hamper the detection of Earth mass planets around such stars.

Results from Doppler spectroscopy

There are very many ground-based telescopes devoted to Doppler spectroscopy aimed at finding exoplanets, and many others that spend part of their time on this quest. The telescopes that collect light for the spectrograph are typically a few meters in aperture. There are also a few telescopes in space that spend at least part of their time searching for exoplanets by Doppler spectroscopy.

For planets down to a few tens of Earth masses, Doppler spectroscopy has been very fruitful, yielding nearly all of the discoveries. It has made a small number of discoveries of planets with masses only a few times that of the Earth, mainly around M dwarfs. F, G, and K dwarfs (the Sun is a G dwarf) are bright enough for Doppler spectroscopy to be used out to thousands of light years, whereas the much fainter M dwarfs need to be considerably closer.

Even though most exoplanets have been discovered by Doppler spectroscopy, observations by transit photometry of the same planet solve the orbital inclination problem, so we know the actual mass of the planet and not just its minimum mass, and we also get its radius. We can then get the volume of the planet, and then its density, which is mass/volume. The density of a planet tells us a lot about its overall composition. This will be discussed further in Chapter 11, which is devoted to the known exoplanetary systems, discovered by all the various methods I have described.

SUMMARY

Given sufficient knowledge about the star, as detailed earlier. Table 10.2 lists what we learn about an exoplanet from the four techniques described in Chapters 9 and 10.

TABLE 10.2 Indirect detection methods compared

	Astrometry	Doppler spectroscopy	Transit photometry	Gravitational microlensing
Mass of planet	yes	minimum value (usually)	no	yes
Radius of planet	no	no	yes	no
Semimajor axis of orbit	yes	from period	from period	projected value
Period of orbit	yes	yes	yes	no
Eccentricity of orbit	yes	yes	no	no

11

The known exoplanetary systems

In this Chapter I'll give you a brief account of the discovery of exoplanetary systems, and then describe the properties of the systems discovered so far. The next chapter considers what sort of exoplanetary systems await discovery. This will raise the question of whether undiscovered rocky-iron planets like the Earth could be present in the classical habitable zones of known exoplanetary systems.

11.1   THE DISCOVERY OF EXOPLANETARY SYSTEMS

In 1992, after decades of disappointment and false hope, the first exoplanets were discovered. The US astronomers Alex Wolszczan and Dale Frail announced that they had detected two planets in orbit around a rare type of star called a pulsar. Each planet had a mass just a few times that of the Earth. The claim that a pulsar had planets was greeted with considerable surprise by other astronomers, but it has withstood further investigation. The surprise stems from the way that a pulsar is formed, as the remnant of a massive star after it explodes in a supernova (Section 7.2). The remnant can either be a black hole or a neutron star. A pulsar is a neutron star that we observe by the beacon of electromagnetic radiation that it sweeps across us as it rotates, giving us a series of regular pulses. The planets were detected through the periodic changes in pulse spacing resulting from the Doppler effect as the pulsar orbited the center of mass of the system. The discovery was thus, like Doppler spectroscopy, based on the Doppler effect, but in a different way.

It had not been anticipated that planets could survive a supernova explosion, and perhaps any pre-existing planets didn't. Instead, the planets could have formed from debris left by the explosion, or maybe they were captured from a companion star as the pulsar travelled near it. More important for us, life could not have survived the explosion, and even if the planets formed afterwards, they would still be uninhabitable so close to the pulsar, with its deadly radiation. I will therefore discount the handful of pulsar planets, and confine my attention to planets around main sequence stars (plus the one planet known to orbit a red giant).

FIGURE 11.1 Michel Mayor and Didier Queloz, discoverers of the first non-pulsar exoplanet, 51 Pegasi b. (Geneva Observatory)

Non-pulsar exoplanets

In the 1990s several ground-based observatories were conducting searches of main sequence stars using Doppler spectroscopy.

It was in October 1995 that two Swiss astronomers, Michel Mayor and Didier Queloz of Geneva Observatory (Figure 11.1), announced the first non-pulsar planet. They made the discovery with the Elodie spectrograph on the 1.93 meter aperture reflecting telescope of the Observatoire de Haute-Provence, which lies at an altitude of 650 meters is south east France. The planet is in orbit around the solar type star 51 Pegasi (star number 51 in the constellation of Pegasus). The result was soon confirmed by others, and by early 1996 astronomers knew that the long drought in exoplanetary discoveries had ended.

A steady trickle of discoveries has followed, and continues today. In March 2008, excluding the four pulsar planets, 273 planets were listed in one catalog, in 236 planetary systems, 26 of which are multiple planet systems (see http://www.exoplanet.eu/catalog.php). In addition there are a hundred or so candidates for planetary status, as yet unconfirmed, mainly from transit and gravitational microlensing surveys. Exoplanets are named after their stars, by using the letter 'b' to denote the first planet in the system to be discovered, 'c' the next, and so on. Thus the planet of 51 Pegasi is named 51 Pegasi b. If several planets in a system are discovered at the same time, then the lettering starts with the innermost and works outwards, as in Upsilon Andromedae b, c, and d.

Box 11.1 Star names

The stars visible to the unaided eye, particularly the brighter ones, have names stretching back into antiquity – Canopus, Polaris, Rigel, Sirius, Vega, and so on. Such stars are also designated by a Greek lower case letter followed by the name of the constellation in which the star appears, roughly in order of decreasing visual brightness, with α (alpha) almost always denoting the brightest star in the constellation, β (beta) almost always denoting the next brightest star, and so on through the Greek alphabet, ending with ω (omega). Thus, Vega, the brightest star in the constellation of Lyra (the harp) is also called α Lyrae. Note that the name changes to the genitive case – Lyra to Lyrae ("lie-ree"). This system has been in operation since it was introduced by the German astronomer Johann Bayer (1572–1625) in the early years of the seventeenth century.

The great majority of stars have no individual names, but the brighter ones are still designated by a Greek letter followed by the constellation name, such as ε (epsilon) Eridani, the fifth brightest star in the constellation Eridanus (the river).

The trouble with the Greek alphabet is that it only has 24 letters, yet there are far more stars in each constellation than this that are visible to the unaided eye, and hugely more that are visible through even a small telescope. There were several attempts to extend the lettering system, but today we use only the Greek letters. To include more stars, numbering systems were introduced, and one still in widespread use was brought in by the British astronomer John Flamsteed (1646–1719). The final version of his catalog, containing nearly 3,000 stars, was published in 1725. Within each constellation, stars are numbered in the order of increasing celestial longitude (called right ascension), as these were in about 1700, and not according to brightness. Examples are 51 Pegasi (star 51 in Pegasus, the flying horse) and 47 Ursae Majoris (star 47 in Ursa Major, The Great Bear).

Also common today is a numbering system published in 1914–1918 by the US astronomer Annie Jump Cannon (1863–1941) and co-workers at Harvard College Observatory. It is named the Henry Draper Catalog after the man whose widow financed the catalog. Stars are again numbered in the order of their celestial longitude in the sky, but are not sorted into constellations, and include fainter stars than in Flamsteed's catalog. There are nearly 360,000 stars in this catalog – HD1, HD2, etc.

Stars also appear in various more specialized star catalogs, for example, there is the Gliese ("glee-za") catalog of all stars within about 245 light years known in 1991. Stars in this catalog are denoted by G or Gl followed by a number. Yet another catalog is devoted to variable stars.

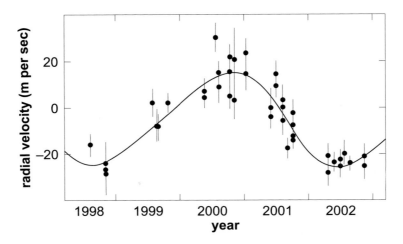

FIGURE 11.2 The variation in radial velocity of the star Tau[1] Gruis (HD216435), at its discovery in 2002. It is about 109 light years away, has a mass about 1.25 times that of the Sun, and might be at the end of its main sequence lifetime. Its planet has a minimum mass 1.49 times that of Jupiter, and moves in an orbit 2.7 AU from its star, with an eccentricity of 0.34.

Almost all of the 273 non-pulsar planets have been discovered using Doppler spectroscopy. The few that have not have been discovered by transit photometry, gravitational microlensing, or direct imaging. Transit photometry and direct imaging give no information about the mass of the orbiting object, but in many cases, planetary masses of the object have subsequently been determined by Doppler spectroscopy.

Figure 10.11 showed radial velocity data for three stars, along with fitted curves. Figure 11.2 shows another example, with a smaller radial velocity range, and a long period. I have also added vertical bars to indicate the uncertainties in the radial velocity measurements. The radial velocity of the center of mass of the system has been subtracted. The data on the star and its only known planet are given in the Figure caption.

The first discovery by another method was in 2003, of a planet orbiting the star OGLE-TR-56. This star was un-named before the discovery, and its name carries that of the survey – OGLE (Optical Gravitational Lensing Experiment, Section 9.3). In spite of the survey name, the discovery was made by transit photometry (Figure 11.3). This is because a survey looking for the light amplification in gravitational microlensing can also detect changes in apparent stellar brightness due to a transit. That OGLE-TR-56 b has planetary mass has been confirmed by Doppler spectroscopy. See the Figure caption for details.

OGLE-TR-56 is about 5,000 light years away, which explains why the light curve is very "noisy" – compare Figure 11.3 with the HST curve for HD209458 in Figure 9.2, which is only about 150 light years away. Some of the few other planets discovered by transit photometry, and all of the few discovered by

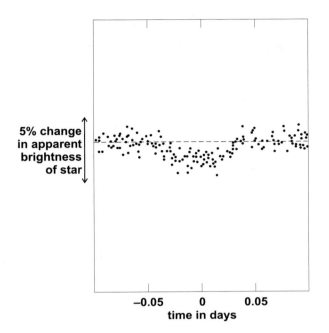

FIGURE 11.3 The light curve of the 1.17 solar mass main sequence star OGLE-TR-56, about 5,000 light years away. The dip is due to its transiting planet. The planet's actual mass is 1.29 times the mass of Jupiter and 1.3 times Jupiter's radius. It orbits at 0.0255 AU with a period of 1.211909 days.

gravitational microlensing, are also far away. The very great majority of the other stars with planets are within 300 light years, which is only about 0.3% of the diameter of the disc of the Galaxy in which we live (Section 7.1). Therefore, all these other exoplanets are in our cosmic backyard. Why so close?

The reason for their proximity is because at close range we receive more stellar radiation, and so all detection methods have an easier time. This is even the case for Doppler spectrometry, even though the Doppler shifts are independent of stellar distance. This is because the closer the star the more distinct are its spectral lines, and the tiny Doppler shifts caused by planetary mass bodies are consequently easier to measure.

Some exoplanets discovered by Doppler spectroscopy have since been detected by transit photometry. Recall that to observe a transit the planet's orbit has to be presented to us very nearly edge-on. Exoplanet orbits are orientated randomly on the sky. Calculations show that in this case, out of the 250 or so *Doppler* discoveries, we can only expect to observe transits for approximately four, which is about what we have.

One system discovered by Doppler spectroscopy, that of the star Gliese 876, has since been detected astrometrically by the HST. This detection was aided by the low mass of Gliese 876, 0.32 solar masses, and its proximity, 15.4 light years, leading to a large astrometric signal. Another astrometric confirmation has been made of the low mass companion of HD89744, by the UK Infrared Telescope on Mauna Kea, but its mass is near the 13 Jupiter mass boundary between giant planets and brown dwarfs, so its planetary status is not confirmed. No exoplanets have yet been *discovered* by astrometry.

Only one exoplanet has so far been detected for certain by imaging, but this orbits the brown dwarf 2M1207 rather than a main sequence star. The planet has a mass roughly five times that of Jupiter, and orbits its star well beyond the HZ. The young K dwarf AB Pic has an object at a projected distance of about 275 AU from its star, but its mass, 13–14 times that of Jupiter, places its planetary status in question. The mass of GQ Lupi b (Section 8.3) is still very uncertain – it could be anywhere in the range 1–40 Jupiter masses.

Let us now look at the discovered exoplanetary systems in more detail, including the stars at their hearts, the planetary orbits, the exoplanets as bodies, and the ways in which the various systems might have formed. Remember that I am excluding the few pulsar planets.

11.2 THE KNOWN (NON-PULSAR) EXOPLANETARY SYSTEMS

I'll first tell you about the *stars* that have planetary systems, and then I'll turn to the planets themselves.

The stars that have planetary systems

Most stars known to have exoplanetary systems lie within a few hundred light years of the Sun – as pointed out in Section 11.1 the nearer the star, the easier it is to make measurements.

The stars that host the known exoplanets are predominantly main sequence stars of spectral type F, G, or K (Section 7.2), that is, they are predominantly stars not very different from the Sun. This is because such stars have attracted most of the search effort by Doppler spectroscopy. They suit this technique because they have plenty of sharp spectral lines, good surface stability, are fairly luminous, and they are not rare, so there are plenty of bright examples. In relation to the search for life it is fortunate that they also have sufficiently long main sequence lifetimes for any life to have effects we might be able to observe (Section 7.3). M dwarfs have even longer lifetimes, and they are also abundant, more so than F, G and K dwarfs put together (Section 7.3). These low luminosity stars must also be included, and the ones near to us are now being scrutinized, with a growing number of successes.

Among the stars known to have planets, several are members of binary systems. In such cases the planet orbits one of the stars. For example, in the binary star Gamma Cephei, a giant planet is in a 2.044 AU orbit around a star 1.6 times the mass of the Sun. The second star, 0.4 solar masses, is at 21.4 AU. It might have been the case that the second star would have prevented planetary formation or ruled out stable orbits, but this is not so. A high proportion of stars are in multiple systems, mainly binary systems – about 70% of stars in the solar neighborhood are in such systems. That many such systems can have planets increases considerably the potential number of stars with planets.

About 50% of the F, G, and K dwarfs within about 200 light years of the Sun have been the targets of Doppler spectroscopy. What proportion of these stars have planets? Certainly much greater than the 10% or so that have had planets detected (by 2007). These are predominantly giant planets. Two incontrovertible selection effects are suppressing the proportion detected. First, some stars are so active that discovery of planets is being masked. Second, some stars have only been observed for times far less than the 10–15 years that are required to find planets in orbits comparable in size to those of Jupiter in our Solar System. Extrapolating from the trends in period in the known systems, it is beyond reasonable doubt that as time passes more planets will be discovered. Just correcting for these two selection effects it is clear that at least 25% of the targets of Doppler spectroscopy have planets, predominantly giants. The proportion that are likely to have planets regardless of mass will be discussed in the next chapter.

To the F, G, and K dwarfs must be added planets that will be discovered around other types of star, particularly the abundant M dwarfs, which, at present, have contributed only a small number of exoplanets.

Nearly all of the stars within about 200 light years have metallicities exceeding 0.5%, some exceeding the Sun's 1.6% (Section 7.2). Such metallicities are comparatively high. This might indicate, reasonably, that planetary formation is more likely with higher proportions of condensable materials in the stellar nebula (Section 2.5), though studies of stars apparently without planets is very incomplete.

Gravitational microlensing surveys reach far out, and the few exoplanets discovered in this way lie thousands of light years away. These surveys favor the direction to the nuclear bulge of the Galaxy (Section 7.1), because the number of stars that can act as background stars makes microlensing a more frequent event than in other directions. On the basis of discoveries to date, it has been estimated that fewer than about a third of the stars in the direction of the bulge have Jupiter mass planets at projected distances in the range 1.5–4 AU from the lensing star, and fewer than half have rather more massive planets in the range 1–7 AU. Note that, because of their abundance, the typical lensing star is an M dwarf.

A transit survey of the stars in the open cluster NGC6819 (Figure 11.4 (top)), which consists of about 150 stars of around solar metallicity, each about 2,400 Myr old, failed to reveal any transits. This was in spite of the expectation of a few, based on statistics from the solar neighborhood. This might just be a statistical fluctuation – bad luck – or indicate that the solar neighborhood is, for some reason, particularly well endowed – NGC6819 is nearly 8,000 light years away.

Equally unsuccessful were surveys for planetary transits among the million or so stars that make up the globular cluster 47 Tucanae (Figure 11.4 (bottom)), and the several million stars that constitute the globular cluster Omega Centauri.

FIGURE 11.4 Top: the open cluster NGC6819 that consists of about 150 stars, each about 2,400 Myr old. (John Mirtle, Calgary, Alberta, Canada). Bottom: the globular cluster 47 Tucanae that consists of about a million ancient stars. (Marcos Mataratzis and Vivek Hira)

However, these ancient stars have low metallicities, which reduces the amount of planet making material, and thus reducing the chance of planets forming. On the other hand, a planet of a few Jupiter masses discovered in the globular cluster M4, in a binary system consisting of a pulsar and a white dwarf, might indicate otherwise. Also, transit surveys favor planets in small orbits, where the orbit is more likely to pass between the star and us, and where the interval between transits is shorter. Therefore planets in large orbits cannot be ruled out in globular clusters. The subject of exoplanets is very young, and there are many uncertainties.

Observational selection effects

You can see that the number of confirmed exoplanets is likely to increase greatly in the next few years. Quite what sort await discovery is the subject of Chapter 12. First, let's examine the ones we *have* discovered. Do bear in mind the selection effects that produce biases in the discoveries. For example, you will readily appreciate that the larger the planet in comparison with its star the more likely it is to be discovered. In Chapters 8–10 the origin of various selection effects were outlined. As well as cropping up in the rest of this chapter, they will be visited again in Section 12.2.

The masses of the known exoplanets

Figure 11.5 shows m_{min} versus the orbital semimajor axis a for each known exoplanet. The mass m_{min} is in units of the mass of Jupiter, denoted by m_J (318 times the mass of the Earth). Recall from Section 10.2 that Doppler spectroscopy gives the mass of the planet were the orbit to be presented edge-on to us. If this is not so then the mass we obtain from measuring radial velocities is the *minimum mass* of the planet, m_{min} – it could be a lot larger if the orbit is presented nearly face-on. Only in a few cases, for example when the planet has also been observed in transit, do we know the orbital inclination and can thus get the actual mass m. If the orientations of the orbits of exoplanets are random, then it can be shown that the *average* value of m is 1.3 m_{min}. It can also be shown that the great majority will have an actual value of m less than 2 m_{min}.

Before we look at the data in Figure 11.5, look at the axes. Equal distances along the axes correspond to equal *multiples*. In moving up the axis showing masses, the marked values, spaced equally, go from 0.01 to 0.1, to 1, to 10. Note also the ticks *between* 0.01 and 0.1. These are at 0.02, 0.03, 0.04, 0.05, 0.06, 0.07, 0.08, 0.09, and similarly between 0.1 and 1.0, and between 1 and 10. This is called a logarithmic scale, and is used where some quantity, here the minimum mass, ranges across widely different values. The a axis is also logarithmic.

From Figure 11.5 you can see that the majority of values of m_{min} lie in the range 0.7-5 m_J, with a fair sprinkling at lower masses. As most of these values have been obtained from Doppler spectroscopy we do not know the actual mass m, but we do know that, overall, were we able to plot m, the distribution in

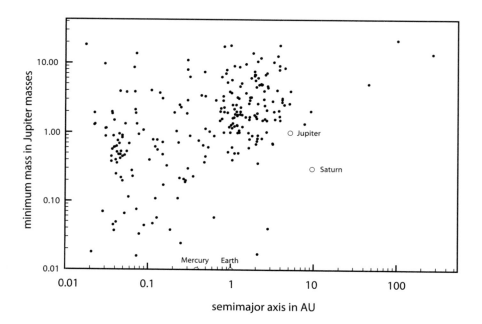

FIGURE 11.5 m_{min} versus orbital semimajor axis a for the known exoplanets. The Earth and Mercury only have their values of a shown – their masses are too small to be shown on this plot. Note the paucity of discoveries in the lower right. This is due to observational selection effects – see the text.

Figure 11.3 will spread a bit, and overall move upwards a bit, with a few outliers off the top of the chart.

The lowest masses

As of March 2008, the least massive (non-pulsar) planet with a confirmed discovery has $m_{min} = 0.0158\ m_J$, which is 5.0 times the mass of the Earth (m_E). This, at 0.073 AU, is the middle planet of the three known to be orbiting the M dwarf Gliese 581, 20.5 light years away. It is Gliese 581 c, and was discovered in April 2007 by Doppler spectroscopy using the HARPS spectrograph on the European Southern Observatory's 3.6 metermeter aperture telescope on La Silla, in Chile.

There are two remarkable things about Gliese 581 c. First, it orbits within the HZ of Gliese 581, which has a mass of only 0.31 solar masses. Second, if it were much more massive than 5.0 m_E then the planetary system of Gliese 581 would be unstable, yet it is about 4,300 million years old – nearly as old as the Solar System. It is thus insufficiently massive to be dominated by hydrogen and helium. It is very likely to be broadly like the Earth in composition, perhaps with more water.

We are thus likely to have found a planet not very different from the Earth, orbiting in the HZ of its star, and about 4,300 million years old. The Gliese 581 c is the first exoplanet discovered that could be habitable.

The second lowest mass planet is Gliese 876 d, the innermost of the three known planets that orbit the M dwarf Gliese 876 (0.32 solar masses). It has m_{min} = 0.018 m_J, which is 5.7 times the mass of the Earth (m_E). It orbits at 0.0208 AU, which is closer to the star than the inner boundary of the HZ. There is some evidence that the orbit is presented not far from edge-on, in which case m is not much greater than m_{min}.

Masses are thus approaching those of the Earth, but we are not there yet, and the great majority of known exoplanets have masses more akin to that of Jupiter.

The highest masses

At the other extreme, we are not concerned with bodies of m_{min} greater than about 13 m_J. Such bodies are better classified as stars than as planets. In particular objects in the mass range 13-80 m_J are the brown dwarfs (possibly down to 10 m_J, thus overlapping with giant planet masses). Those above about 80 m_J are stars proper (Section 7.2). There seems to be a paucity of bodies with m_{min} around 13 m_J, which reduces the number of borderline candidates for planethood. As I pointed out above, if we were able to plot actual masses, this will not much increase the population around 13 m_J. This also means that few of the objects in Figure 11.5 are brown dwarfs rather than planets.

The (non pulsar) exoplanets known at present thus have m_{min} ranging from 0.0158 m_J to 13 m_J, with a lot in the range 0.7-5 m_J. But do these planets also resemble Jupiter in composition? And what of the less massive and more massive exoplanets – what are they made of? I'll return to these questions in Section 11.3.

Exoplanet orbits – minimum masses versus semimajor axes

A striking feature of Figure 11.5 is how small many of the semimajor axes are. You can see that some are over ten times smaller than that of Mercury, the planet closest to the Sun in the Solar System. This proximity would be much less remarkable if the exoplanets were low mass rocky-iron bodies, as in the Solar System, but their masses are too high for this – they are hydrogen-helium giants. To see why proximity is then remarkable you need to recall from Section 2.4 that giant planets are widely thought to have formed beyond the ice line, which would have been at about 4 AU from the Sun when it was young. Therefore, if, as seems very likely, the giant exoplanets formed no closer than several AU, then those giants that are now well within the ice line of their star must have subsequently moved inwards. These planets thus indicate planetary migration. That theoreticians have found reasonably convincing migration mechanisms that can produce giants close in, adds weight to the view that these exoplanets, like all Jupiter mass exoplanets, are predominantly hydrogen-helium in composition. I'll return to migration in Section 11.4.

At the other extreme, there are, as yet, few exoplanets at and beyond Jupiter's distance from the Sun, and all but a couple of these have minimum masses that are over twice the mass of Jupiter. Consequently, the lower right of Figure 11.5 is rather empty. This is due to the observational selection effects whereby

- the more massive a planet the easier it is to detect by Doppler spectroscopy, and, as high mass goes with large size, by transit photometry; and
- the further the planet is from the star the longer its period, and the longer we need to observe it to get data over a few orbits.

The empty quarter in Figure 11.5 is gradually shrinking as stars are observed for longer, and with more sensitive equipment.

Exoplanet orbits – orbital eccentricities versus semimajor axes

Figure 11.6 shows the eccentricities e versus semimajor axis a, of the known exoplanets, with Jupiter and Saturn for comparison (note that the e axis is linear, not logarithmic). Some of these eccentricities are very high, particularly at the larger values of a (Figure 2.2 shows an orbit with $e = 0.7057$). It is difficult to produce such large values by formation from a circumstellar disc, without something extra. One possibility is close encounters between giant planets. Simulations show that a common outcome is for one giant to be flung into the depths of interstellar space and for the other to be retained in a high e orbit.

Another possibility arises when the orbital periods of the two giants are in a simple ratio, such as 1:2 or 1:3. In such a case the gravitational interaction

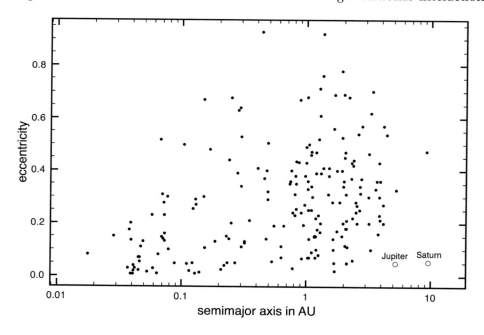

FIGURE 11.6 Eccentricities e versus semimajor axis a for the known exoplanets.

between the giant planets can disturb the orbits in a cumulative way, resulting in large orbital changes. An everyday analog is the pushing of a child on a swing – if you time your pushes correctly a large amplitude builds up. Such simple ratios can arise through the migration of giant planets (Section 11.4).

These scenarios for generating large values of e are also the basis for an alternative explanation of giants close to the star. If the surviving giant is in a high e orbit with a small periastron distance (the distance when it is closest to the star), then tidal interactions with the star, perhaps aided by residual disc gas, will reduce e and we can end up with a giant in a small, low e orbit. If the other giant avoided ejection it will be in a large, high e orbit.

In conclusion, we certainly have plausible explanations of the orbits of exoplanets. The giant planets in the Solar System emerge as just one of many possible outcomes.

11.3   SIZES AND COMPOSITIONS OF EXOPLANETS

You have seen that the actual mass of an exoplanet can be obtained from its minimum mass if it is also observed in transit – we then know the inclination of its orbit with respect to the plane of the sky (it will be close to 90°, Section 10.2).

Suppose that we know the actual mass of an exoplanet to be equal to the mass of Jupiter, and therefore 318 times the mass of the Earth, but that we know nothing else about it. What could we say about its composition? Almost nothing. We would expect it to consist mainly of the most abundant chemical elements in the Galaxy – hydrogen, helium, carbon, nitrogen, oxygen, magnesium, silicon, and iron. But which of these dominate? If the Jupiter-mass planet resembles Jupiter in composition then hydrogen and helium will dominate; if it resembles a massive Earth, then it will be dominated by silicon, oxygen, magnesium, and iron – it will be a rocky-iron body.

We can eliminate one of these possibilities by obtaining the average density of the planet – its mass divided by its volume i.e.,

$$\text{density} = \text{mass/volume}$$

The average density of Jupiter is 1,330 kg per cubic meter, and that of the Earth is 5,520 kg per cubic meter (liquid water at the Earth's surface has a density of about 1,000 kg per cubic meter). There is thus a clear difference in the densities. The difference would be even greater if the self-compression in a massive planet were not more than in a low mass planet. The average density can thus tell us something about the composition.

In order to get the average density we clearly need to know the volume of a planet. This can be obtained from its radius, which in turn can be obtained if it is observed in transit. The apparent decrease in brightness of the star during transit gives us A_p/A_{star}, where A_p is the cross sectional area of the planet, and A_{star} is the cross sectional area of the star (Section 9.1). The value of A_{star} can be obtained through a variety of observations of the star. For us, the important point is that if we know A_{star} then we get A_p, and from A_p we get the radius of the planet.

Box 11.2 Getting the radius of an exoplanet

Planets are near enough spherical, and therefore for a planet with a radius R_p the volume V_p is:

$$V_p = \frac{4}{3} \pi R_p^3$$

Where π ("pie") is the circumference of a circle divided by its diameter. To three figures it is 3.14, but it is a curious number, one that goes on without end, 3.141 592 635...... It is an example of a transcendental number.

In order to get V_p we need to get R_p. This is obtained from the cross sectional area A_p of a planet, which is obtained from transit photometry. The equation is:

$$A_p = \pi R_p^2$$

Therefore:

$$R_p = \sqrt{\frac{A_p}{\pi}}$$

and so V_p is obtained from the first equation.

What sort of results have been obtained? Consider the first transit, observed in 1999, of the planet orbiting the solar type star HD209458 (Section 9.1). This planet, HD209458 b, had been discovered earlier that year by Doppler spectroscopy, which had determined its minimum mass to be 0.69 m_J, where m_J is the mass of Jupiter. The transit turned this into its actual mass, and also established its radius to be 1.46 R_J, where R_J is the radius of Jupiter. This was later revised downwards to 1.30 R_J. Thus, HD209458 b is 2.2 times the volume of Jupiter (1.30^3), yet it is only 0.69 m_J. Its density is thus 420 kg per cubic meter, 0.31 times that of Jupiter. With Jupiter dominated by the two lightest elements, hydrogen and helium, what can HD209458 b be made of?

It is surely also dominated by hydrogen and helium. The main reason for its large size is its proximity to its star, only 0.045 AU – the Earth is 1 AU from the Sun. You might think that thermal expansion of its atmosphere bloated it, but this is a minor effect. More important is the reduction in the rate of cooling of the planet, due to the heat of its star. If the planet has been close to its star from soon after its formation then we have a plausible explanation of its bloated size today – it has not contracted as much as an otherwise similar planet would have further from the star. There might be other contributions too, notably tidal heating by the star.

About 15 other planets detected in transit have had their masses determined by Doppler spectroscopy. The masses cover the approximate range 0.33–1.3 mJ, and the densities 300–1,500 kg per cubic meter, with little correlation between the two. None are far from their star, and so are likely to be bloated. All are surely dominated by hydrogen and helium, though not necessarily all to the same extent.

For the other exoplanets we have only indirect evidence for their compositions. It is unlikely that any of the giant planets are other than hydrogen-helium dominated. For example, for a Jupiter mass planet to have a rocky-iron composition like the Earth the circumstellar discs from which they formed must have had very high metallicities. The metallicities are unrealistically large given what we know about the composition of the sort of interstellar clouds from which stars and their planetary systems have been born. In any case, such enrichment would be seen as very high metallicity in the parent stars. The stellar spectra do indeed indicate some enrichment, but nowhere near large enough to support the hypothesis that the giant exoplanets broadly resemble the Earth in composition.

There is thus good evidence that the giant exoplanets so far discovered have compositions that resemble Jupiter in that they are dominated by hydrogen and helium. But how far down in mass can this be a reasonable conclusion? Saturn, which is 0.30 times the mass of Jupiter, 95 times the mass of the Earth, is also dominated by hydrogen and helium (Section 2.3). But Neptune, at 17 times the mass of the Earth ($0.054\ m_J$), is dominated by icy and rocky materials (Section 2.3). At these lowish masses the distance from the star also matters – distant planets of lowish mass are likely to be icy-rocky, whereas those near the star will be rocky-iron.

The lowest dozen minimum masses in Figure 11.5 are less than the mass of Neptune. Most of these are close to their star (it being easier to detect planets there), and therefore they are likely to be rocky-iron – "super-Earths"!

Detection of radiation from exoplanetary atmospheres

A small number of exoplanets have had some atmospheric constituents identified. A splendid example is the transiting planet HD 209458 b. In 2001 the HST detected sodium vapor in its atmosphere. In 2003–4 an enormous envelope of hydrogen, carbon, and oxygen was observed. Gases are escaping from the envelope, though it is thought that only about 7% of the mass of the planet will be lost over the estimated 5,000 Myr main sequence lifetime of the star.

In 2005 infrared radiation emitted by the planet was detected by the Spitzer Space Telescope, the first time this had been achieved for an exoplanet. The detection was made by comparing the infrared radiation received from the star when the planet was not in transit, with that when it was. From the measured planetary infrared radiation the temperature of the region in the atmosphere radiating to space was calculated to be 750°C. This is not surprising given that HD209458 b is only 0.045 AU from its type G main sequence star.

Infrared spectra of HD209458 b were obtained by Spitzer in 2007, over the wavelength range 7.5-13.2 micrometers. These spectra contained some surprises, notably the initial failure to detect water vapor, and the detection of what could be silicate dust. There was also a peak in the emission at 7.78 micrometers, as yet unexplained. HST observations were re-examined later in 2007, and it seems possible that water vapor is present in the atmosphere of HD209458 b, though this is not yet confirmed.

Spitzer has also observed HD189733 b, a 1.15 mJ planet that transits a 0.8 solar mass star at a distance of 0.0312 AU. The results are broadly the same as for HD209458 b – atmospheric temperatures around 700°C, and good evidence for water, this time without the aid of the HST.

Note that detection in situ of what must surely be the dominant atmospheric constituents, hydrogen and helium, is not yet possible because these gases have weak infrared signatures. It is the low density of the planet that tells us that they dominate the overall composition.

HD209458 b and HD189733 b are typical of what are called hot Jupiters, an apt name given their proximity to their stars. Recall that it is believed that giant planets formed further away, and migrated inwards. Let us now examine the extraordinary likelihood of the migration of giant planets.

11.4 MIGRATION OF GIANT EXOPLANETS

The first few (non-pulsar) exoplanets to be discovered were giants in small orbits, in a few cases astonishingly so. Table 11.1 lists the first eight to be discovered, all by Doppler spectroscopy, all of them orbiting stars not very different from the Sun.

TABLE 11.1 The first eight exoplanets to be discovered (ordered by semimajor axis).

Stellar characteristics[1]		Planetary characteristics[1]					
Name	Mass (solar masses)	Distance in light years	Orbital period in days	Semimajor axis in AU	Eccentricity	Minimum mass (Jupiter masses)	Year of discovery
Tau Bootis	1.25	49	3.312	0.042	0.0	3.64	1996
51 Pegasi	1.0	50.2	4.229	0.051	0.01	0.47	1995
Upsilon Andromedae	1.25	53.8	4.620	0.053	0.03	0.63	1996
55 Cancri A	0.84	43.7	14.66	0.11	0.05	0.84	1996
Rho Coronae Borealis	1.0	54.4	39.8	0.23	0.028	1.1	1997
70 Virginis	0.95	59.0	116.7	0.47	0.40	6.84	1996
16 Cygni B	1.0	72	802.8	1.70	0.68	1.74	1996
47 Ursae Majoris	1.1	46.0	1093	2.08	0.09	2.42	1996

(1) The values are at discovery. Some of the data have since been revised significantly, particularly for three of these systems where further planets have been discovered: 55 Cancri A (5 planets), Ups And (3 planets), 47 UMa (2 planets).

All the planets in Table 11.1 are much closer to their star that the distance at which they must have formed. This must have been beyond the ice line i.e., that distance from the star beyond which water condensed to provide much of the mass of the icy-rocky kernels from which the giant planets grew. This was by capturing gas (mainly hydrogen and helium) from the stellar nebula that gave birth to each exoplanetary system, much as the solar nebula gave birth to the Solar System (Section 2.5).

Those within about 0.05 AU of F, G, and K dwarfs were soon called hot Jupiters. In my opinion this term could be extended to those further away – recall that Mercury, the planet closest to the Sun, is a scorched world in an orbit with a = 0.387 AU. But I'll adopt convention, and call giant planets in the range 0.05-0.4 AU from F, G, and K dwarfs, *warm* Jupiters.

Within a few months of the first discoveries, plausible mechanisms were suggested by which planets could migrate inwards to lie closer to the star than the ice line, even to become warm or hot Jupiters. Theoreticians were not being entirely "wise after the event" – migration had been predicted over a decade earlier, in research that was largely overlooked.

Migration mechanisms, and consequences for giant planets

The key to migration is the gravitational effect of the giant planet on the circumstellar disc of gas and dust in which it is embedded and from which it has formed. The details are complex so only a qualitative outline is given here.

At first the disc is symmetrical about an axis running perpendicularly through the protostar at its center. But as the mass of the embryonic giant grows, its gravity produces spiral structures in the disc that destroy this symmetry, as in Figure 11.7 (note that the disc is not modeled close to the star, hence the "black hole", which is an artefact). Consider the formation of a giant planet at the point where its icy-rocky kernel has a mass less than that of the Earth. The spiral structure in the disc interior to the kernel has a gravitational influence on the kernel's orbit tending to push it outwards, whereas the spiral structure exterior to the disc exerts a gravitational influence that tends to push it inwards. For any plausible disc model the inwards push is the greater and so the net effect is inward migration. The rate of migration is proportional to the mass of the disc and also to the mass of the kernel, so as it grows it migrates inwards ever more rapidly. This is called Type I migration. Note that the disc itself is also migrating inwards, but always more slowly than the kernel.

Type I migration continues until the kernel has grown to sufficient mass to open up a gap in the disc, as illustrated in Figure 11.8. The gap causes a major change in the migration. It now slows dramatically, by 10–100 times, until the kernel and the disc are migrating inwards at the same rate. This is Type II migration. The kernel mass at which the transition takes place depends on various properties of the disc (its density, thickness, viscosity, temperature, and so on), and on the distance of the kernel from the star. An approximate range is 10–100 Earth masses, and so it is likely that there is a fully fledged giant planet

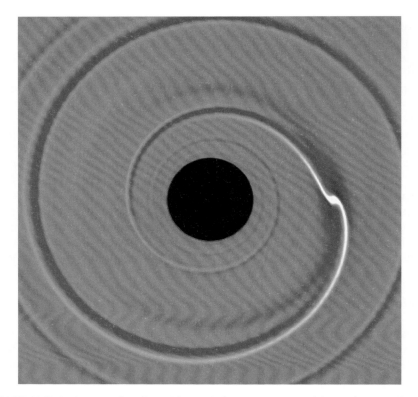

FIGURE 11.7 A circumstellar disc with a spiral structure created by a planetary kernel. The disc is not modeled close to the star, hence the "black hole", which is an artefact. (F Masset 2004)

kernel of icy and rocky materials, plus some captured gas, at this transition. Gap formation reduces the rate at which the kernel acquires mass from the disc, but does not halt it, and so planets up to several Jupiter masses can form.

Migration needs to be halted, otherwise all the giant planets in every system will end up in the star. Either the disc must be removed, or there must be counter effects. The disc will be removed partly through gradual infall to the star and partly through bursts of activity that young stars are observed to undergo, when outflowing stellar winds and intense UV radiation somehow dissipate the disc. This is the T-Tauri phase. Observations of young stars suggest that the disc lasts 1–10 Ma. This will be too long in some cases for the giant planet to survive, because the Type II migration time, depending on the disc properties and other parameters, can be less than this. There are however several ways in which counter effects can appear, including

- tidal interactions between the planet and the star;
- the effect of mass loss from distended young giant planets, which act as "rockets";

FIGURE 11.8 The kernel of a giant planet opens up a gap in its circumstellar disc. The "black hole" is an artefact. The filaments from the kernel are feeding disc gas onto it. (F Masset 2004)

- magnetic interactions between the star and the disc that halt the inward migration of the disc, and consequently halts Type II migration too, because in Type II migration the planet and disc migrate together; and
- evaporation by stellar radiation of a narrow zone in the dish shaped disc at a few AU, thus creating a barrier.

These mechanisms (and others) are rather subtle and are peripheral to the story, so they are not detailed here, but note that they could save otherwise doomed planets.

So far we have considered a sole giant planet in a disc, yet the Solar System and at least several of the known exoplanetary systems have more than one giant, and it is presumed that most of the other systems do too. Computer models of discs with two giants have shown that interactions between the giants can slow and even reverse Type II migration, and this is one way in which we could have ended up with a Solar System in which Jupiter and Saturn are still beyond the ice line. It is also possible under special choices of circumstellar stellar disc parameters to have limited migration without invoking giant–giant interactions.

Another migration mechanism is possible after the disc has cleared, and this is through the scattering of planetesimals by giant planets. In this way an inner giant planet can migrate inwards, and an outer giant outwards. This seems to have happened in the Solar System.

Giant planets closer to the star than the ice line thus have ready explanations, through migration.

But what about rocky-iron planets in the HZ? Could these be present even after a giant planet has crashed through the HZ? If they cannot be present, then the proportion of known exoplanetary systems that could be inhabited is much reduced.

11.5 CONSEQUENCES OF MIGRATION IN THE CLASSICAL HABITABLE ZONE

Of particular interest is rocky-iron planets with masses greater than about 0.3 m_E, with a likely upper limit of a few m_E, the upper limit being determined by the availability of materials. I shall call such planets "Earth-type". Take care to distinguish this term from "Earth mass", which is self-explanatory, and Earth-twin, a planet the size of the Earth, and, like the Earth, 1 AU from a solar-type star. As discussed in Section 7.4, if, what we are now calling an Earth-type planet, is in the HZ it could have a habitable surface.

The speedy Type I migration would rapidly carry Earth-type planets from the HZ into the star. Only by a narrow choice of the circumstellar disc parameters is it possible for such planets to migrate sufficiently slowly to outlast the disc and thus survive. This can seem rather contrived. Fortunately, models reveal that the growth of Earth-type planets is likely to be rather slow, much slower than the growth of giant planet kernels beyond the ice line. During the few million year lifetime of the circumstellar disc, the bodies interior to the ice line will only have acquired masses up to about a tenth of the mass of the Earth – these are called embryos. The migration of such small bodies will have been slight. There will also be smaller bodies, called planetesimals (Section 2.4).

Meanwhile, growth has been more rapid beyond the ice line, because the abundance of condensable water provides a swarm of planetesimals with low relative speeds, which can thus coalesce to form kernels. Subsequently, as you have seen, the region interior to the ice line might or might not be traversed by a (growing) giant planet. If the inward migration of the giant is, at most, slight, objects in the HZ will not be greatly disturbed by migration. In other systems, with more extensive migration, even if the giant(s) stops short of the HZ, embryos and planetesimals will be scattered as the giant moves inwards, sweeping orbital resonances across the HZ. If the HZ is actually *traversed* by a migrating giant then embryos will certainly be scattered. So, could Earth-type planets form after migration has ceased, and could any such planets be present in the HZ today?

Formation of Earth-type planets in the HZ

The few computer studies so far made have shown it to be possible for Earth-type planets to form in the HZ after the migration of a giant planet through this zone to become a hot Jupiter, well interior to the HZ. Hot Jupiters do not subsequently threaten the survival of Earth-type planets in the HZ. Some planetesimals and embryos will have been captured by the giant, others will have collide with the star, and some will have been flung out of the system. But many survive, and the remnant gas in the disc helps to circularize the orbits, particularly of the smaller bodies.

Post-migration formation of Earth-type planets ensues. Additionally, there could be sufficient dust in the disc to form a new generation of planetesimals that leads to a new generation of embryos. The chance of at least one Earth-type planet forming in the HZ is reduced compared to cases when the giant planets remain well outside the HZ, like Jupiter in the Solar System, but it is far from negligible.

Less well studied is the case where a giant planet migrates towards the HZ but stops outside its outer boundary. This will have the effect of scattering the planetesimals and embryos, and will surely reduce the probability of formation of an Earth-type planet, but the details are unknown.

Another effect of the inward migration of a giant planet is to shepherd inwards ice rich bodies from beyond the ice line, perhaps beyond embryo size. This would increase considerably the amount of icy material available in the formation of Earth-type planets. This raises the possibility that any Earth-type planets that form post-migration could be several times the mass of the Earth, and covered in deep oceans of water.

11.6   SURVIVAL OF EARTH-TYPE PLANETS IN THE CLASSICAL HABITABLE ZONE

If, in any system, an Earth-type planet somehow formed in the HZ, the next question is whether it could have been there for at least the past 1,000 million or so years, making due allowance for the outward migration of the HZ, and excluding the first 700 Myr of its life to allow for a presumed heavy bombardment (Section 2.4). This is not the same question as that of formation of an Earth, which takes no more than about 100 Ma. By contrast, survival is against the gravitational buffeting by the giant planet(s) for the present age of the star, which is typically several billion years.

These time spans are based on the Earth, where it took about 2,000 Myr after its origin before life had an effect on the atmosphere that could be detected from afar – mainly the oxygen generated by photosynthesis (Section 5.4).

Computer studies have been made, by me and others, of nearly all of the known exoplanetary systems, to see whether an Earth-type planet could be present today in each system and have been in the HZ for at least the past 1000 Myr (subsequent to the first 700 Myr of its lifetime). On the basis of these studies it can be concluded that in perhaps a half of the known exoplanetary systems, Earth-type planets, provided always that they did form, meet this requirement. In some cases the planet could have survived anywhere in the HZ, and in other cases only in variously restricted regions. If, in the studies, an Earth-type planet has been removed, its usual fate was to be flung into the cold of interstellar space, or to have a collision with its star.

If formation of an Earth-type planet were not possible in a system with a giant planet interior to the HZ, then only about 7% of the known exoplanetary systems would have Earth-type planets in the HZ.

Figure 11.9 shows four of various configurations of the HZ and a giant planet. Figure 11.9(a) shows a giant planet in a low eccentricity orbit near its star, well interior to HZ – a hot Jupiter. An Earth-type planet could be present anywhere within the HZ (provided that it formed). Figure 11.9(b) shows a system like ours,

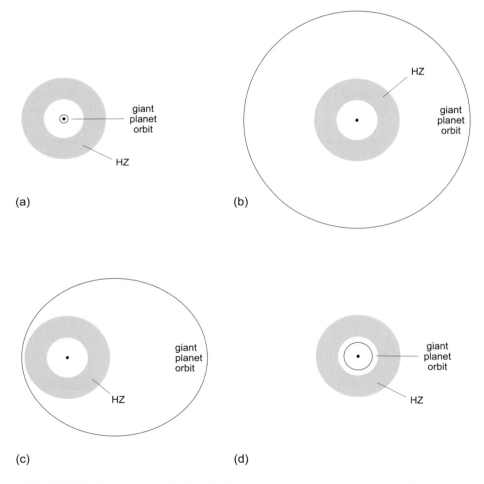

FIGURE 11.9 Survivable orbits for Earth-type planets in the classical habitable zone (shaded) where (a) the giant planet is a hot Jupiter (b) the giant is well outside the HZ (as is the case of Jupiter in the Solar System) (c) the giant is not far outside the HZ (d) the giant planet is not very far interior to the HZ.

where the giant planet Jupiter is well beyond the HZ. In such systems the whole HZ, like ours, would be a safe harbor for Earth-type planets. There are, as yet, no other known systems like ours, though they very probably await discovery (see Section 12.2). Figure 11.9(c) shows a giant not far outside the HZ. Such cases need individual scrutiny. Earth-type planets might be present only in the inner region of the HZ. Figure 11.9(d) shows a giant planet not very far interior to the HZ. Such cases also need individual scrutiny – Earth-type planets might be present only in the outer region of the HZ. As well as these four cases other cases occur, including those in which the giant planet in Figure 11.9(b) has a very eccentric orbit.

Remember that the HZ moves outwards during the main sequence lifetime of a star, as its luminosity slowly increases (Section 6.1). Two examples are shown in Figure 11.10. For stars like the Sun this can leave planets near the inner edge of the HZ behind, and incorporate new planets initially beyond the HZ. This has been taken into account in the above statistics derived from computer models. In the case of lower mass stars, such as the M dwarf in Figure 11.10, the HZ has hardly moved – this is because of their very long main sequence lifetimes.

The other place where Earth-type planetary bodies could be in a safe harbor is as massive satellites of giant planets that orbit within the HZ. One example is HD23079, a main sequence star with a mass of about 1.1 solar masses, and about 3,000 Myr old. Its giant planet, minimum mass 2.54 m_J, is in a low eccentricity orbit with $a = 1.48$ AU, which puts it in the middle of the present day HZ. It is not known whether it has an Earth-type satellite. If such satellites are common, then this would increase significantly the number of exoplanetary systems that could be habitable, even inhabited.

Let's now turn to the exoplanetary systems that still await discovery.

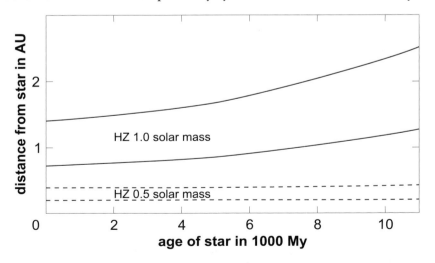

FIGURE 11.10 The classical habitable zones of two main sequence stars: an M dwarf with a mass half that of the Sun, and a star of solar mass. (Figure 7.6 is essentially the same.)

Box 11.3 The planetary system of 55 Cancri

As of March 2008 the 1.03 solar mass dwarf star 55 Cancri A has the greatest number of known exoplanets, five. It is also in a binary star system. Its companion, 55 Cancri B, is a red dwarf over 1,000 AU away. The planetary masses and orbits are as follows.

Planet	Semimajor axis in AU	Orbital eccentricity	Minimum mass (Jupiter masses)
e	0.038	0.07	0.034
b	0.115	0.014	0.824
c	0.240	0.086	0.169
f	0.781	0.2 approx.	0.144
d	5.77	0.025	3.835

The illustration shows the inner four planets and also the inner three planets in the Solar System for comparison. The fifth planet, 55 Cancri Ad, is in a much larger orbit. All five planets are giants. You can see that 55 Cancri Ae is a hot Jupiter, though "hot Neptune" would be more apt!

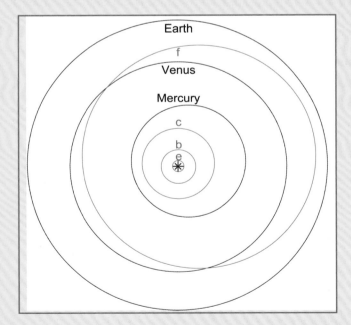

The orbits of the inner four planets of 55 Cancri A compared to the orbits of Mercury, Venus, and the Earth.

12

The undiscovered exoplanetary systems

What sort of exoplanetary systems await discovery? Are we on the brink of discovering planets like the Earth orbiting in the classical habitable zone, and thus habitable, even inhabited? First, I'll summarize the material covered previously (mainly in Chapter 11) about the exoplanetary systems that have been discovered so far.

12.1 THE KNOWN EXOPLANETARY SYSTEMS – A SUMMARY

As of March 2008, Doppler spectroscopy has made the very great majority of the 273 confirmed discoveries of non-pulsar planets, in 236 planetary systems, 26 of which are known to be multiple planet systems (see http://www.exoplanet.eu/catalog.php). A significant and increasing number has been discovered by transit photometry and gravitational microlensing. No exoplanets have yet been *discovered* by astrometry, though one system, that of the star Gliese 876, has been detected astrometrically after its discovery by Doppler spectroscopy. None has been discovered for certain by direct imaging, though the case of GQLupi b is still under review.

The properties of the known exoplanetary systems can be summarized as follows.

- The most distant planets have been discovered by gravitational microlensing, and they orbit stars thousands of light years away.
- The very great majority of other stars with planets are within a few hundred light years of the Sun.
- The nearest star with a planet is Epsilon Eridani, at 10.4 light years. It is orbited at 3.39 AU by a giant planet with an actual mass 1.55 times that of Jupiter.
- At least 25% of the F, G, or K main sequence stars within about two hundred light years of the Sun have planets.
- Stars with high metallicity seem to be favored for having planetary systems.
- The planets range in minimum mass from 0.0158 Jupiter masses (5.0 Earth masses) to the brown dwarf limit, 13 Jupiter masses, with a preponderance in the range 0.7–5 Jupiter masses.

- The giant planets are thought to be rich in hydrogen and helium, though only for a few planets is there evidence for this, from their mean densities.
- Roughly 40% of the known exoplanets orbit their star more closely than Mercury orbits the Sun, and about 5% orbit beyond 4 AU.
- The great majority of exoplanets have orbital eccentricities greater than that of Jupiter (0.0489), in many cases much greater, particularly in the larger orbits (the greatest value is 0.927).
- A few stars with planets are in binary stellar systems, the planet(s) orbiting just one of the two stars.

Gravitational lensing surveys have shown that fewer than about a third of the stars that lie in the direction of the nuclear bulge of the Galaxy have Jupiter mass planets at projected distances in the range 1.5–4 AU from the star, and fewer than about a half have rather more massive planets in the range 1–7 AU. Note that, because of their abundance, the typical lensing star is an M dwarf. A transit survey failed to detect exoplanets in the open cluster NGC6819, in spite of expectations based on extrapolation from the proportions in the solar neighborhood. The reasons for this are uncertain. A survey of the global clusters 47 Tucanae and Omega Centuari also failed to find any planets, perhaps because the stars have low metallicity. In the globular cluster M4 just one planet has been discovered, with a mass of a few Jupiter masses.

12.2   WHAT SORT OF PLANETS AWAIT DISCOVERY?

Exoplanets continue to be discovered, and it is clear that we have not discovered all of them, not even those within a few hundred light years of the Sun. We have not even acquired a representative sample. In particular, observational selection effects are playing a major role, and the consequences of these will be detailed below.

I will be referring to the various types of planet that have been defined in earlier Chapters. Here is a summary.

- The terms "Earth mass", "Earth size", "Jupiter mass", "Jupiter size" are self-explanatory.
- An Earth-*type* planet is a rocky-iron planet having a mass between about 0.3 times of the mass of the Earth, and several times the Earth's mass.
- An Earth-*twin* is an Earth-mass planet orbiting a solar-type star about the same age as the Sun, with a semimajor axis around 1 AU i.e., in the classical habitable zone (HZ).
- A Jupiter-twin is a Jupiter-mass planet orbiting a solar-type, solar age star in an orbit with a semimajor axis of about 5.2 AU.

Remember that a solar-type star is a main sequence star of spectral class G i.e., a G dwarf.

Not just Earth-type planets, but all the "Earths" above will be rocky-iron in

FIGURE 12.1 An artist's impression of a planet that is far too hot to be habitable. (NASA/ JPL-CalTech)

composition, though perhaps in some cases, including Earth-type, with a greater proportion of their mass made of volatile materials, particularly water. Earth-twins are likely to be habitable (even inhabited). This is also the case for Earth mass, Earth size and Earth-type too if they orbit in the HZ. A Jupiter-twin will be dominated by hydrogen and helium.

I will examine at what distance from us planets will be discovered, what masses they will have, what radii, and in what orbits. I'll then outline the stars that will feature in the discoveries. When I'm discussing a particular property, all other relevant properties must be regarded as fixed. For example, in the discussion of planetary mass, it is assumed that the mass of the main sequence star it orbits and its distance from us are fixed, and that the orbit of the planet is fixed in size and in inclination.

At this point, if you feel the need, you should review the five search techniques described in Chapters 8–10.

Stellar distances

Regardless of the technique used to search for exoplanets, in all of them the nearer the exoplanetary system is to us, the easier it is to discover. This is even the case for Doppler spectroscopy. Even though the Doppler shift in the star's spectral lines is independent of the star's distance, we still have to detect the shift – the nearer the star the greater the radiation flux at our telescopes, the easier it is to detect this shift, and the more accurately it can be measured.

Thus, with improved instrumentation, all techniques will discover exoplanetary systems further and further away. As our reach increases, the number of stars increases rapidly, so a cornucopia of discoveries can be expected within a few decades.

Planetary masses and radii

Astrometry, Doppler spectroscopy, and gravitational microlensing, all benefit from an exoplanet having a large mass. In astrometry, the larger the mass of the planet the greater the orbital motion of the star; in Doppler spectroscopy, as planetary mass increases the faster the star moves around its orbit and so the greater the Doppler shift; gravitational microlensing benefits from larger mass because of the longer duration of the planet's effect on the light curve, which makes it more likely to be detected.

Transit photometry gives a bigger signal the larger the radius of the planet, because during a transit the reduction in apparent brightness of the star is greater. Imaging at visible and near infrared wavelengths also benefits from larger radius, because of the associated greater brightness of the planet (assuming it has the same reflectance).

Clearly, in the future, with improvements in instruments, we expect to discover smaller planets, be this in size or in mass, though transit photometry is limited by the masking effect of stellar variability to a few tenths of the Earth's radius.

Planetary orbits

The larger the orbit of a planet the easier it will be to detect by imaging. Also, the size of the star's orbit (around the center of mass of the system) will increase, which facilitates detection by astrometry. Improvements in instruments will thus discover planets in *smaller* orbits in the future.

For Doppler spectroscopy the reverse is the case. The larger the planetary orbit the greater its orbital period, and the greater the orbital period of its star – these periods are the same. The Doppler shift is then less. Moreover, a star has to be observed for longer to accumulate measurements around an appreciable portion of its orbit. In future, as instruments improve, and as observation times increase, planets in *larger* orbits than hitherto will be discovered.

In the case of transit photometry, the larger the orbit the less likely it is that transits will occur, and if they do occur, they will occur less often. However, longer observation times will discover transits that are as yet undiscovered.

Gravitational microlensing has no simple relationship with the size of the planet's orbit. For a lensing star of solar mass, and typical distances to the lensing star and to the background star, the size of the Einstein ring is a few AU. This needs to be the distance from the planet to the lensing star projected on the sky. Future discoveries are likely to be more of the same, in terms of planetary orbits.

The stars in exoplanetary systems

F, G, and K dwarfs have been the focus of attention, so it is not surprising that most exoplanets are in orbit around such stars. However, the abundant M dwarfs are now getting the attention they deserve, though because they are of low luminosity, considerable improvements in the sensitivity of the various techniques are necessary before exoplanets orbiting M dwarfs are detectable as far off as is presently possible for F, G, and K dwarfs. Such improvements will be made. These, plus the increased attention that M dwarfs are getting, means that we can expect, as time goes by, M dwarfs to account for an increasing proportion of stars known to have planets.

Because of the abundance of M dwarfs, it is they that are the typical lensing star in gravitational microlensing discoveries of planets, and this will remain the case.

A summary diagram

Figure 12.2 is my attempt to summarize this section so far. It is a plot of the mass, or radius, of an exoplanet, versus its semimajor axis, or orbital period. The mass of its star and the star's distance from us, are fixed across the diagram. The five techniques that I have described are shown as labels on various parts of the diagram. The arrows show what sort of planets are likely to become discoverable with each technique, as the instruments improve.

You can see that there are no numbers on the plot. It is a qualitative representation, indicating the expected trends in future discoveries.

Figure 12.2 shows that we expect the future to deliver exoplanets of lower mass/smaller radius. The proportion of planets in the total exoplanet population significantly less massive/smaller than Jupiter is thus expected to increase, in fact down to planets less massive/smaller than the Earth.

The proportion of planets in orbits with larger semimajor axes/orbital periods is also expected to increase, despite the increasing proportion with *smaller*

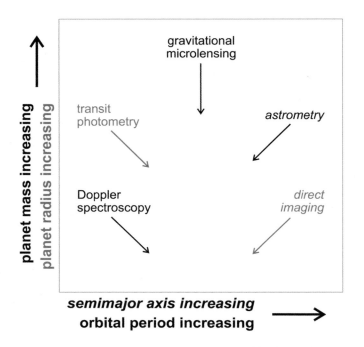

FIGURE 12.2 The sort of planets likely to be discovered by five techniques as they are improved. The techniques in red go with the red label on the *y*-axis. On the *x*-axis the italicized label goes with the italicized techniques.

semimajor axes/orbital periods expected to be discovered by astrometry and by direct imaging. This is because Doppler spectroscopy, at least for the next decade or so, is expected to remain the most fruitful technique, so dominates the trend.

A cautionary tale

In Section 8.3 I mentioned that in the three years up to 2007 a survey was made of 54 young, nearby stars that were among the best candidates for having detectable giant planets in orbits *beyond* 5 AU (Jupiter orbits at 5.2 AU from the Sun). Such far flung planets are easier to see than those closer to the star, and in a young system will be warmer and thus emit more infrared radiation than older planets. The survey was performed with one of the 8.2 meter telescopes of the VLT, and the 6.5 meter MMT in Arizona. The search was made at infrared wavelengths where the spectral features of methane lie – this gas is expected to be fairly abundant in the atmospheres of giant planets. No planets were detected.

This absence of giant planets at the sort of distances at which Jupiter and Saturn lie in our Solar System is striking. Are planetary systems with giant planets beyond about 5 AU rare? This question cannot yet be answered – further surveys are needed.

FIGURE 12.3 An Earth-type planet in an exoplanetary system. The planet is warm enough to have oceans of liquid water. (Julian Baum, Take 27 Ltd.)

12.3   EARTH-TYPE PLANETS

Recall that an Earth-type planet is a rocky-iron planet having a mass somewhere from about 0.3 times the mass of the Earth, up to several times the Earth's mass. Such planets are expected to have an endowment of volatile substances to form an ocean and atmosphere. Figure 12.3 is an artist's impression of one such planet.

Our capacity to discover Earth-type planets is steadily increasing, However, we will, of course, only detect them if they are there. Recall from Section 11.6 that computer studies have shown the following.

- In about half of the known exoplanetary systems, Earth-type planets, *provided that they could have formed in the HZ region*, could have survived in at least some parts of the HZ for at least the past 1,000 Myr (subsequent to a presumed heavy bombardment in the first 700 Myr of a planet's life). This is long enough for any life to have had an effect we could observe, if the Earth is any guide.
- The HZ of the rest of the known exoplanetary systems does not offer a safe haven to Earth-type planets, because the giant planets are close enough to destroy their orbital stability throughout the HZ.
- If Earth-type planets *cannot* form in systems where a giant planet orbits

between the star and the inner boundary of the HZ, and if the giant got there by migration through the HZ, then only about 7% of the known exoplanetary systems could have Earth-type planets in the HZ. However, evidence for formation being prevented by migration is weak. Migration of giant planets makes it less likely that Earth-type planets could form in the HZ, though perhaps not much less likely.

Combining these points with evidence from further computer studies that shows that if an Earth-type planet *could* form in the HZ then it *will* form there, we can expect to find Earth-type planets in the HZ of about half of the known exoplanetary systems. We can also expect to find such planets interior to the inner boundary of the HZ, such as analogs of Venus in the Solar System, or beyond the HZ, where the Solar System has never had an Earth-type planet (Mars has a mass only about a tenth of that of the Earth).

You know that at least 25% of the F, G, or K main sequence stars within a few hundred light years of the Sun are known to have planets. This proportion is a lower limit for main sequence stars, given that observational selection effects have discriminated against M dwarfs, which comprise about 80% of the stars, against Earth-type planets, and against systems like ours where the giant planets are well beyond the HZ.

Surveys of nearby M dwarfs with Doppler spectroscopy have established that they are about five times less likely to have giant planets than more massive main sequence stars. However, this does not mean that M dwarfs are less likely to have Earth-type planets – the paucity of giant planets could well be due to the low mass of the circumstellar gas and dust disc surrounding M dwarfs during their formation. Moreover, the absence of giant planets would reduce the chance of the HZ being disturbed gravitationally. On the other hand, the discovery of Earth-type planets close to M dwarfs by Doppler spectroscopy might be hampered though stellar activity (Section 10.2).

Nevertheless, within 400 hundred light years of the Sun there are roughly half a million stars. Habitable exoplanets might, therefore, not be far away.

To summarize: within a decade we expect the proportion of stars known to have planets to rise considerably. Among these new discoveries there should be lower mass planets, down to Earth mass, or even less, and planets of all masses in larger orbits. By the 2020s we will have space telescopes such as Darwin and TPF (I or C) capable of seeing the planets themselves, even Earth-size planets. Huge ground-based telescopes such as E-ELT might also have this capability in favorable cases.

At present the Solar System looks rare, with its giant planets in large, low eccentricity orbits, and its whole HZ consequently a particularly secure abode for the formation and long-term survival of Earth-type planets. We expect systems more resembling the Solar System to be discovered, and perhaps even to be common. By the 2020s we should know how common they really are. Then we can enjoy 20:20 vision!

12.4 EVIDENCE FROM CIRCUMSTELLAR DISCS

Among the planets awaiting discovery, it is expected that some will be around those stars known to be surrounded by gas and dust. Many young stars are known to have discs of gas plus some dust, and a few older stars are known to have discs or rings of dust, the gas having been dissipated by stellar activity, notably during the T Tauri phase.

As noted earlier, Epsilon Eridani is known to have a planet – it also has a dust ring. The planet orbits inside the ring and might be too far from it to cause gravitational distortions in the ring. The ring however is not uniform, and the cause might be planets near its inner boundary. Distortion is also present in the dust disc around Beta Pictoris (Figure 12.4). The disc is presented to us nearly edge-on, and it is clearly warped. This warp could well be due to the gravitational influence of a giant planet.

Epsilon Eridani and Beta Pictoris are just two examples. Other dust discs have central holes that indicate removal of dust by planet formation. Around older main sequence stars, the very existence today of dusty discs or rings calls for a mechanism to replenish the dust that otherwise would have been long gone. A likely mechanism is collisions between asteroids, perhaps comets too in an

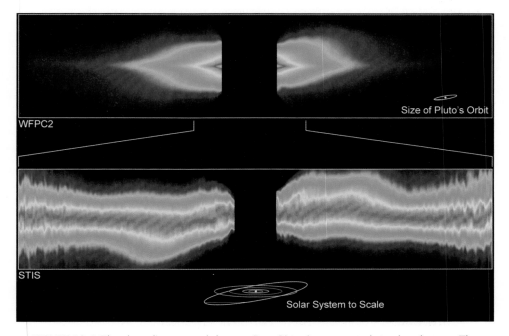

FIGURE 12.4 The dust disc around the star Beta Pictoris, presented nearly edge-on. The density decreases above and below the mid-plane. The gap in the middle of the lower image is part of the imaging system, and is about 30 AU across. (PRC98-03, STScI/NASA Sally Heap (GSFC) and Al Schultz (CSC and STScI))

analog of the Edgeworth–Kuiper belt in the Solar System (Section 2.4). If there are such bodies, there may be planets too. However, from a survey of a small sample of F, G, and K main sequence stars, it seems that few stars possess a substantial dust disc *and* giant planets orbiting at less than a few AU from the star. If this is generally the case, then the discovery of dust discs on its own does not imply the presence of planets relatively close to the star.

That completes my review of the exoplanetary systems, known, and awaiting discovery. Next I'll look forward to that time when we have not only discovered Earth-type planets, but have obtained images, even if only as a dot of light, from which we can see whether they could bear life, and indeed whether they do so.

13

Finding life on explanets

Now we come to the heart of the matter. Imagine a time, in the not too distant future, when we can observe a dot of light that is an exoplanet. We would much prefer to see the planet as a disc, but that possibility lies further in the future. So, what could we do with a dot of light in our search for life?

In this chapter I'll concentrate on the detection of life based on complex carbon compounds and liquid water i.e., on carbon-liquid water life. I'll thus concentrate on life that resembles life on Earth. But I'm not assuming that alien life is based on the *same* carbon compounds as terrestrial life. It might use a carbon compound other than DNA to carry genetic information, and it might use carbon compounds other than proteins to carry out the various functions performed by proteins in terrestrial life. But it would still be carbon-liquid water life. In Chapters 14 and 15 I will free us of this restriction. In Chapter 14, the search for extraterrestrial intelligence (SETI), is a search for function, for evidence of technological civilization, regardless of the chemical basis of the organisms. In Chapter 15 I will consider alternative biochemistries.

It might seem very parochial to concentrate on carbon-liquid water life. One justification is that this is the nature of the only life we know to exist. Another justification stems from fundamental chemistry. No other element has anywhere near the same facility as carbon to form compounds of sufficient complexity, diversity, and versatility to support the many processes of life. Few liquids approach water in their usefulness as solvents and reactants. A third justification is that we know how to detect evidence of carbon-liquid water life. Apart from SETI, we have a far poorer idea of how to detect life that has an entirely different chemical basis from ours.

We will be particularly concerned with life that we could detect from afar, which probably restricts us to the surfaces of planets that have been in the classical habitable zone (HZ) of their star for at least the past 1,000 Myr or so (subsequent to any heavy bombardment – Section 2.5). Using the Earth as a guide, this should be long enough for life to have had an effect on a planet's surface or atmosphere that we could detect. We will not be much concerned here with life outside the HZ, where it would probably be confined to the interiors of planetary bodies, though such life might somehow produce distantly observable effects.

You will see that the external physical form of alien life is not relevant to this chapter. Only in Chapter 15 do I present speculations on possible physical forms of aliens.

13.1   PLANETS WITH HABITABLE SURFACES

To be habitable, a planet (or a satellite) must meet conditions involving its orbit, composition, and mass. These conditions were discussed in Section 7.4, but are summarized here.

- For *surface* habitability, it must orbit in the HZ, so that any water present would be stable as a liquid over a substantial part of its surface.
- It must have a solid surface, even if it is overlain by a deep ocean. Therefore, the planet must be predominantly rocky. It must also have volatiles that include water. The volatiles must form an atmosphere sufficient to keep surface water stable as a liquid. There must also be carbon compounds.
- The planet could also be an icy-rocky body that formed beyond the ice line and migrated or was shepherded inwards. In the HZ the icy materials would melt, and being predominantly water, would form a deep global ocean.
- To retain a substantial atmosphere the mass must not be less than about 0.3 Earth masses. The upper mass limit of a rocky planet is a few times the mass of the Earth, as determined by the likely amounts of rocky materials (including iron), or 10–20 times the mass of the Earth for an icy-rocky body.

There are yet further requirements for habitability. For example, if a giant planet orbits in or near the HZ, habitable planets will be ejected from it (Section 11.6). As another example, if there are too many large impacts, then life might never develop, so the rate of impacts that could cause mass extinctions, even global sterilization, needs to be below some threshold. For these reasons (among others) a proportion of otherwise habitable planets will be uninhabitable, though it is not possible on present knowledge to put a figure on this. On the other hand a giant planet in the HZ could have a large, habitable satellite (Figure 13.1).

To establish whether an exoplanet is *habitable* is the first step. The second is to see whether an exoplanet (or a satellite of a giant exoplanet) is not just a potential habitat but an actual habitat. For each step we need to be able to analyze the electromagnetic radiation we get from that dot of light that is an exoplanet, and this requires sufficient spatial resolution to isolate the planet's radiation from that of the star. This will be possible in the next decade when the various new ground-based and space telescopes described in Chapter 8 become available. Much further into the future it will be possible to send probes to the stars. But whether our instruments are on a probe approaching a planet, or orbiting the Sun or the Earth, or lie at the Earth's surface, we need to ask – how could remote observations find life out there?

FIGURE 13.1 A habitable satellite of a giant planet in the classical habitable zone. (Julian Baum, Take 27 Ltd.)

13.2   IS THERE LIFE ON EARTH?

In 1989 the Galileo Orbiter was launched by NASA (Figure 13.2). Its primary mission was to study Jupiter and its satellites. It reached Jupiter in December 1995, but before that, in December 1990 and again in December 1992, it came within about 6 million km of the Earth. This was in order to gain kinetic energy through gravitational interaction with the Earth, thus enabling Galileo's modest rockets to raise the massive payload to far away Jupiter. Advantage was taken of these close encounters to see whether life could be detected on the Earth and on the Moon. In 1990 the Earth was the object of study, and in 1992 the Moon. Even though the instruments on board were not designed specifically to detect life, it was hoped that the outcome would help astronomers to design ways of detecting life from afar.

The answer that the Galileo Orbiter gave to the question: "Is there life on Earth?" is a resounding "Yes!" The conclusion itself came as no surprise, but it was encouraging that Galileo's instruments could give it. There were three instruments that provided the evidence.

Evidence from infrared absorption

First, there was the infrared spectrometer called NIMS. This enabled astronomers to identify atmospheric substances through the absorption imprint they placed

FIGURE 13.2 The Galileo spacecraft, *en route* to Jupiter – an artist's impression. (NASA/ CalTech)

on the spectrum of infrared radiation *emitted* by the Earth. The substances detected included ozone (O_3) and methane (CH_4). Recall from Section 5.4 that one of the major effects of the Earth's biosphere is that it sustains molecular oxygen (O_2) as a major component of our atmosphere. O_2 has only a weak spectral signature at infrared wavelengths, but through the action of solar UV radiation, O_2 gives rise to an appreciable trace of O_3, and this has such a strong spectral signature in the infrared that it is readily detected.

It is difficult to envisage any process other than photosynthesis that could generate sufficient O_2 to yield the amount of O_3 seen. But we can't be quite sure of this. The clincher is CH_4. It very readily reacts with O_2 to give CO_2 and H_2O, and as a result CH_4 only accounts for about 1 molecule in every 600,000 in the Earth's lower atmosphere. This however was sufficient to give NIMS a clear if small infrared signature. The crucial point is that without a huge rate of release of CH_4 into the oxygen-rich atmosphere, the quantity of atmospheric CH_4 would be many orders of magnitude less, and it would have been undetectable by NIMS. Non-biological sources, such as volcanoes, are insufficient. The huge rate of release comes from the biosphere. The presence of O_2 and CH_4 together, far from chemical equilibrium with each other, puts the existence of a biosphere on Earth beyond reasonable doubt.

Evidence from the reflection of solar radiation

Second, instruments on the Galileo Orbiter measured the amount of solar radiation that the Earth *reflected* at various wavelengths – a reflection spectrum. Around a wavelength of 0.8 micrometers, which is in the near infrared, a sharp rise in reflectance was detected, particularly over the continents. The 'red edge' as it is called, is due to green vegetation, and is associated with the photosynthesizing molecule chlorophyll and with biological structures that reject radiation not utilized by chlorophyll.

At red and blue wavelengths there is low reflectance, due to absorption by chlorophyll – it is this absorbed energy that drives photosynthesis. Red light is utilized because many red photons reach the Earth's surface, and blue light because each photon carries a lot of energy due to its short wavelength (Section 2.2), and can thus drive certain photochemical reactions that less energetic photons could not. Green photons fall between two stools – less abundant than red photons but each is not as energetic as a blue photon. Consequently, the absorption of green photons is weak – green light is reflected away, which is why plants on Earth look green.

However, though we know how to interpret these spectral features on Earth, it is not at all certain that photosynthesis in an alien biosphere would look the same, particularly if the spectrum of radiation from its star differed significantly from that of sunlight, as it would, for example, for an M star. Perhaps the best that can be hoped for is to see absorption that could not be readily accounted for by common minerals. Overall, this is a less certain indicator of life than pairs of atmospheric gases way out of chemical equilibrium, such as O_2 and CH_4. Moreover, a biosphere need not contain any photosynthesizing organisms – life is not contingent on this, as you saw in Section 3.3.

Evidence from the emission of radio waves

Third, the radio receiver on Galileo detected strong radiation confined to a set of very narrow wavelength ranges. Moreover, the radiation at each of these wavelengths was not constant but was modulated in an intricate way that could not be explained by natural processes. What do you think these were?

They were the various terrestrial radio and television transmissions. The modulation was the information that carried the program content – a steady wave carries no soap operas. This shows that the Earth not only has a biosphere, but that it has evolved in a particular, possibly very rare manner, to yield a technological civilization. Unambiguous images of cities and other artefacts were not obtained by the relatively small Galileo cameras at the rather large distances of this spacecraft. Figure 13.3(a) shows the modest resolution achieved. Figure 13.3(b) shows the far superior resolution achieved by a satellite orbiting close to the Earth, but even in this case cities and artefacts cannot be seen, though distinct areas of green vegetation can be seen in Turkey.

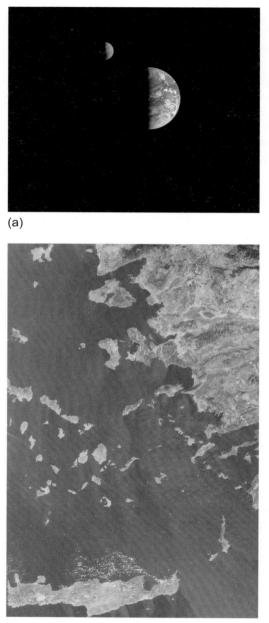

(a)

(b)

FIGURE 13.3 (a) The Galileo spacecraft views the Earth and Moon, 16 December 1992, from a range of 6.2 million km. (NASA/JPL) (b) NASA's Terra satellite views the eastern Mediterranean in September 2001. The area covered is a few hundred kilometers across. (JPL/NASA- CalTech, PIA03725)

In December 1992 Galileo's instruments were turned to face the Moon, with entirely negative results! There were no pairs of atmospheric gases way out of chemical equilibrium, indeed, hardly any gases at all, no characteristic red edge, no radio or TV broadcasts. If there is life on the Moon it must be deep in the crust, and this is extremely unlikely given the scarcity of water.

We will now look at infrared and visible/near infrared spectra in more detail, deferring to Chapter 14 the detection of radio transmissions. Spectra are of huge importance because they could reveal a biosphere at a great distance from us. Consider first infrared spectra.

13.3   THE INFRARED SPECTRUM OF THE EARTH

Figure 13.4 shows the infrared emission spectrum of the Earth, as seen from space. This is not the spectrum obtained by the Galileo spacecraft but a much more detailed one obtained by the Nimbus-4 satellite in the 1970s. This particular spectrum was acquired in daytime above the western Pacific Ocean, and has been chosen because it resembles the sort of spectrum that would be obtained from a cloud-free Earth from a great distance, when the light from the whole planet would enter the spectrometer. The vertical scale shows the power emitted from the Earth at each wavelength (note that the wavelength scale is logarithmic – see Section 11.2).

There are several smooth curves, each labelled with a temperature. These correspond to emission from a surface that is black at infrared wavelengths, and has temperatures equal to those shown. There is also a jagged curve displaying much detail. This is the infrared power emitted by the Earth. It is the detail in

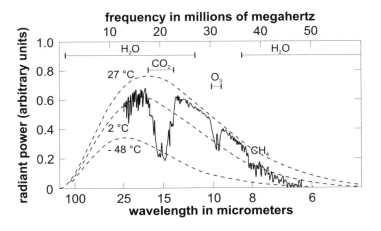

FIGURE 13.4 Earth's infrared emission spectrum, as obtained in daytime by the Nimbus-4 satellite over a cloud-free part of the western Pacific Ocean in the 1970s. (Note that as wavelength decreases the frequency rises, which is true for all waves.)

this curve that provides the evidence that the Earth is inhabited. Let's examine Figure 13.4.

Surface temperature (and pressure)

Between 12 micrometers and 8 micrometers, except for the dip around 9.6 micrometers, the terrestrial spectrum follows closely the curve labelled 27°C. The absence of deep, narrow spectral absorption features, such as would be generated by abundant atmospheric gases, indicates that the radiation received by Nimbus-4 in this wavelength range has been emitted by the Earth's surface or by clouds. It has in fact been emitted by the surface, the area being cloud free. The average temperature of the (daytime) surface at this particular time over the western Pacific Ocean is close to 27°C, perhaps a little less.

At this temperature, if the atmospheric pressure is sufficiently high, water can exist as a liquid at the surface. Water exists as a liquid from about 0°C to a higher temperature depending on the atmospheric pressure. The pressure can be estimated from various external measurements, including the width of spectral lines (I'll not go into details). At the surface of the Earth such measurement would reveal that this pressure enables water to be liquid up to about 100°C. So we can conclude that liquid water could exist at the Earth's surface.

For an exoplanet we would not know whether clouds or the surface was responsible for the 8–12 micrometer spectrum. If the cloud cover were variable, or if there were spectral features that could be linked to cloud particles, then probably we could tell.

The other important inference from a surface temperature of 27°C is that it is well within the range for complex carbon compounds to exist. Compounds such as proteins and DNA break up at temperatures above about 160°C, so most of the Earth's surface is safely cool. On the other hand it is not so cold that biochemical reaction rates, which decrease rapidly as temperature falls, are too low to sustain life.

Water

That water is *actually* present on Earth, at least as vapor in the atmosphere, is indicated by much of the fine structure in the spectrum in Figure 13.4. The H_2O molecule has a great many narrow absorption lines in the infrared spectrum, so many that they overlap and blend together to form absorption bands. It is these bands rather than the individual lines that are seen in abundance in Figure 13.4. More evidence for water is at the ends of the wavelength range. The smooth curve with a temperature of 2°C fits the spectrum fairly well, but this temperature is too low for the surface of the western Pacific Ocean. At these wavelengths, the water vapor has absorbed all the radiation emitted from the surface, and has re-emitted it at the lower temperature at its higher location.

We can thus infer that the atmospheric temperature at the general altitude of the water vapor is lower than at the surface. The actual altitude corresponding to

2°C cannot be obtained from Figure 13.4. To obtain it we need the variation of atmospheric temperature with altitude above the Earth, and this is shown in Figure 13.5, obtained from direct measurements and averaged over the Earth's surface. You can see that 2°C occurs at an altitude of only a few kilometers. Therefore, there must be enough water vapor below this altitude to hide the ground from space at the extreme wavelengths. This is in accord with direct measurements that show water vapor to be heavily concentrated in the lower few kilometers of the Earth's atmosphere.

Carbon dioxide

At around 15 micrometers the Earth's spectrum in Figure 13.4 has a large dip corresponding to a deep and broad set of overlapping absorption bands due to carbon dioxide, CO_2. The heaviest absorption corresponds to a temperature of about –50°C, and Figure 13.5 shows that this value occurs in the upper

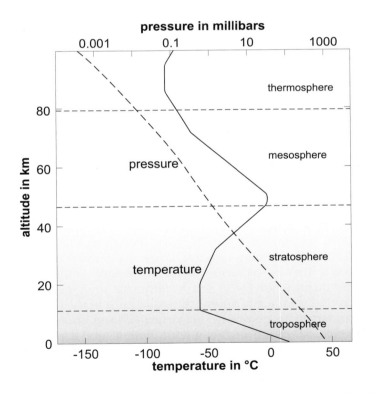

FIGURE 13.5 The variation of atmospheric temperature and pressure with altitude above the Earth's surface. This is a model based on measurements. The angular nature of the temperature graph is a feature of the model – in reality the graph would be smooth. (Note the logarithmic scale for pressure at the top of the chart.)

troposphere. This is not because CO_2 is concentrated there – it is not – but because this is as deep as we can see at 15 micrometers into an atmosphere in which there is sufficient CO_2 at deeper levels to screen them from view. From Figure 13.4 we can conclude that the Earth has carbon, which is essential for biomolecules. The actual amount of CO_2 is about 0.037% (as a proportion of all molecules).

Oxygen

The dip around 9.6 micrometers in Figure 13.4 is due to ozone O_3. This is derived from O_2 by the action of solar UV radiation, and the size of the O_3 feature in Figure 13.4 shows that O_2 must be present in considerable quantity in the Earth's atmosphere. O_2 is not directly detectable in the infrared because it absorbs infrared radiation very weakly. This is because it consists of two identical atoms. In such molecules, roughly speaking, the distribution of electric charge is symmetrical around its center. As a result, infrared radiation cannot exert much net force on the molecule and so the absorption of photons in the infrared is weak. The O_3 molecule has a different symmetry from O_2, and the net force is correspondingly stronger. It is also stronger in molecules consisting of more than one element, diatomic (two atoms) or otherwise, such as CO, CO_2, and H_2O, and so these also have strong infrared absorption.

The temperature around the center of the O_3 absorption is about 0°C. Therefore, Figure 13.5 places the O_3 that is radiating directly to space either in the lower troposphere or in the upper stratosphere. In fact, O_3 is concentrated in the stratosphere so it is the upper stratosphere from where the infrared radiation in Figure 13.4 is coming. The solar UV radiation thus creates O_3 well above the Earth's surface, and it screens the lower atmosphere from the UV that has produced the O_3.

The other question that arises from the presence of O_3 is the source of the O_2 that gave it birth – is it a biosphere? A non-biogenic origin is the photodissociation of water (Section 5.4), in which a UV photon splits H_2O into OH and H. The hydrogen, being of low mass, is lost to space. Reactions of OH lead to the formation of O_2. Photodissociation of water has always been generating some O_2 on Earth. However, the O_2 oxidizes surface rocks and volcanic gases, and consequently the quantity of O_2 in the atmosphere from photodissociation alone would be *far* less than it is. This indicates that most of the O_2 in the Earth's atmosphere is sustained biogenically, in particular through oxygenic photosynthesis. A significant O_3 absorption can result from a good deal less O_2 than is present, but probably not from the small quantity sustained today by photodissociation of H_2O.

There are, however, two conditions under which the photodissociation of H_2O could give rise to O_2 in abundance without the aid of a biosphere. First, it could do so if the rate of photodissociation were far higher. This will be the case in a few 1,000 Myr when the luminosity of the Sun will have increased to the point where the Earth's upper atmosphere will be warm enough to hold a lot more water vapor than it does now. More water means more photodissociation means more O_2. However, this enhancement cannot last long. Photodissociation destroys water, and in a few million years all the water will be lost, and the Earth will be dry. In a comparable time the O_2 produced by photodissociation will be removed through oxidation of surface rocks and volcanic gases. Thus, unless the Earth was caught in the act of losing its water, photodissociation could not account for the high H_2O abundance. Venus seems to have already lost its water in this way – this would have happened early in its history because it is closer to the Sun than is the Earth.

Second, the present rate of photodissociation of water could give rise to a lot of O_2 if it were being removed geologically at a *very* low rate. It would then build up slowly over hundreds of millions of years. In this regard, size matters. Large rocky planets like the Earth sustain considerable geological activity (with plate tectonics at its heart in the Earth's case – Section 2.3). Small planets are unlikely to sustain such activity. Therefore, if a small planet is sufficiently near its star so that its atmosphere is always damp, it could gradually build up a lot of O_2 from the photodissociation of water. However, a geologically inactive planet would be unable to keep its atmosphere damp, or even retain much of an atmosphere at all (Section 7.4)! Any atmosphere rich in O_2 would thus be short lived.

Thus, for the Earth, a combination of size and the unlikelihood of catching our planet at the moment it loses nearly all its water, would make it seem likely to an observer in space that oxygenic photosynthesis was at work.

Methane (and oxygen)

In Section 13.2 it was mentioned that CH_4 is also produced by the biosphere. In Figure 13.4 the tiny trace in the atmosphere is nevertheless sufficient to create absorption lines near 8 micrometers. It is generated by large organisms, notably by bacteria in the guts of ruminants, by certain bacteria elsewhere, in marshes (methane is also called marsh gas), and by paddy fields. As noted in Section 13.2, it is the presence of O_2 and CH_4 together, far from chemical equilibrium with each other, which would put the existence of a biosphere on Earth beyond reasonable doubt to an alien scrutinizing our infrared emission spectrum.

13.4   THE INFRARED SPECTRUM OF MARS

Figure 13.6 shows the infrared spectrum of Mars, at mid-latitudes in daytime under clear conditions, obtained by an orbiting spacecraft, Mariner 9. The surface temperature is about 7°C, chilly, but no threat to the biochemicals of life. The CO_2 absorption feature is clear, but there is no evidence of O_3, and so O_2 must be present as a trace at most. This is indeed the case. There is therefore no evidence for a biosphere in Figure 13.6. Nevertheless, there could be one that has created no atmospheric signature. Indeed, until O_2 was present in substantial quantities in the Earth's atmosphere from about 2,300 Ma ago (Section 5.4) it

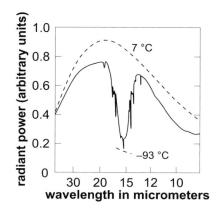

FIGURE 13.6 The infrared emission spectrum of Mars, at mid-latitudes in daytime under clear conditions, obtained by the orbiting spacecraft Mariner 9 in the early 1970s.

would have been difficult to find convincing evidence of life on Earth from its infrared spectrum. However, there is evidence from landers that there is no life on the Martian *surface* today (Section 6.2).

What about water? There is no evidence for it in the spectrum in Figure 13.6, and indeed the atmosphere of Mars is almost devoid of water vapor.

Therefore, if there is life on Mars it must be deep under the surface. Indeed this is where one might still find liquid water continuously present, in contrast to the surface, where water might only appear for brief periods, if at all (Section 6.2).

13.5 THE INFRARED SPECTRA OF EXOPLANETS

Now, to that exciting time when a telescope has obtained an image of an Earth-type planet in an exoplanetary system. Recall that until we send our instruments into an exoplanetary system this will be no more than a dot or smudge of light, with no discernible disc or surface features. But, dot or disc, the light can be passed into an infrared spectrometer, and, as you have seen, can therefore be analyzed for signs of life. Figure 13.7 shows the sort of spectrum that might be obtained by the Darwin space telescope (Section 8.4).

The first thing to notice is that the spectrum is far coarser than those shown in Figures 13.4 and 13.6. To acquire the fine detail seen in those spectra we would have to accumulate radiation for far, far longer than the notional 40 day exposure for the spectrum in Figure 13.7.

In spite of the low resolution, we can discern CO_2 and O_3 absorptions. Furthermore, the reduced temperature in the spectrum at 6–8 micrometers is suggestive of water vapor in the atmosphere. From this coarse spectrum it is not feasible to infer the atmospheric pressure at the surface, and therefore, in the absence of other measurements, we could not be certain that water would be

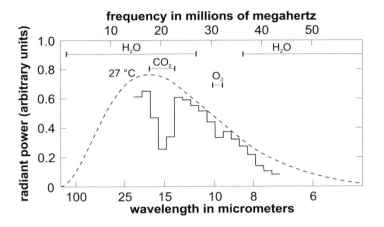

FIGURE 13.7 The infrared emission spectrum that might be obtained by a Darwin sized infrared space telescope, with a 40 day exposure of an Earth-type planet in the classical habitable zone of a star about 30 light years away. Note how coarse the spectrum is.

stable as a liquid at the surface, but if the mass of the planet exceeds about a third that of the Earth, this is likely (Section 7.4). Any CH_4 absorption is beyond detection.

We can estimate the surface temperature from the spectrum at 8–12 micrometers. In this particular case it is about the same as that in Figure 13.4, for the Earth – about 27°C. Except for the possibility that the O_3 has a non-biogenic origin, the concurrence of water, carbon dioxide and ozone, and a suitable surface temperature, indicates that there could be a biosphere, and that within it there are organisms that generate oxygen through photosynthesis.

If only things could be so clear cut!

The case of no oxygen

Suppose that the exoplanet spectrum resembled that in Figure 13.7 *except* that the O_3 absorption is absent. We are now in the tantalizing position that the planet seems *habitable* – presence of carbon and water, the right surface temperature over much of its surface – but there is no evidence that it is *inhabited*.

At this point we must recall an important dictum: "Absence of evidence is not evidence of absence." There are several reasons why there could be a biosphere but no detectable oxygen. It could be that there is plenty of O_2 but too little UV from the star to form O_3. Alternatively, the O_3 might be efficiently removed in some way. But even if there really is no O_2 present the planet could still be inhabited. There are at least three possibilities.

- There *is* a surface biosphere that includes oxygenic photosynthesis, but it is in the state that the biosphere was on Earth at the earliest times – it has not been able to build the oxygen content sufficiently for detection from afar.

- There *is* a surface biosphere, but it either performs photosynthesis in a manner that releases no oxygen, or it does not rely on photosynthesis at all. There are terrestrial organisms that act in each of these ways.
- The biosphere is deep in the crust and does not influence the atmosphere, as might be the case on Mars today, or under an icy crust such as on Europa.

But in spite of these possibilities, there are other ways in which we could detect evidence of life in the infrared spectrum. An important possibility is that we might detect in the infrared spectrum gases other than O_3 that defied explanation by non-biological processes.

However, whereas one such gas would be suggestive, we must always be aware of plausible non biological explanations, such as the photodissociation of water that yields the O_2 that leads to O_3. Two gases that should readily react to virtually eliminate the presence of one or both of them would be a much stronger indication of life. In the case of the Earth one pair is O_2 and CH_4 – you have seen that the trace of CH_4 has been driven by biochemical production to be far more abundant than it would if it were in chemical equilibrium with O_2. If any out of equilibrium pair of gases was identified, then strenuous attempts would be made to find plausible non-biological explanations, but if all of these failed, then there would be a good case for an alien biochemistry.

13.6   EXOPLANET SPECTRA AT VISIBLE AND NEAR INFRARED WAVELENGTHS

The infrared spectrum of an exoplanet over the wavelength range in Figure 13.7 is generated by emission from the planet's surface and atmosphere, and by the subsequent absorption of some of these emissions by atmospheric constituents. But you have seen in my account of observations of the Earth by the Galileo spacecraft in Section 13.2 that there are other ways in which we might detect life. This section is concerned with observations at visible and near infrared wavelengths.

Evidence of photosynthesis

The most readily detectable spectral feature of our biosphere at these wavelengths is the red edge at near infrared wavelengths around 0.8 micrometers, produced by green vegetation. This is shown in Figure 13.8. This is a *reflectance* spectrum. Take care not to confuse it with the *emission* spectra we have been considering before. In a reflectance spectrum the vertical axis shows the fraction of the radiation from some separate radiation source (such as the Sun) that is being reflected by the planet. In an emission spectrum the radiation originates from the surface of the planet being examined.

You have seen in Section 13.2 that the red edge is associated with chlorophyll, the molecule in green plants (and in some unicellular organisms) that is central

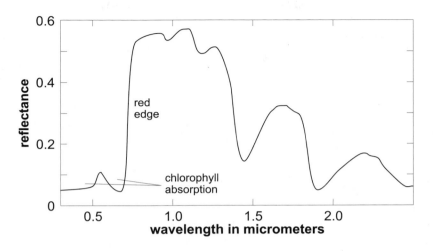

FIGURE 13.8 The reflectance spectrum of a deciduous leaf.

to photosynthesis. You also saw that the absorption at red and blue wavelengths, which results in the green color of vegetation, is also due to chlorophyll.

But there are some terrestrial organisms that utilize different wavelengths for photosynthesis. In some cases this is an adaptation to the solar spectrum as modified by their environments, such as underwater, where red light is attenuated. Photosynthesizers on exoplanets will presumably adapt to the spectrum delivered to them by their star. For example, if red light was not as useful as adjacent wavelengths, plants would mostly look red, and there would be lower reflectance at the adjacent wavelengths than at red wavelengths.

Wherever in the visible-near infrared spectrum the features from photosynthesis lie, large optical telescopes or large interferometers will be required to reveal them. It will be at least 10 years before we have this capability.

Evidence from light curves

There is another way we could use the reflection from an exoplanet to search for life, and this will also be feasible for the forthcoming generation of large telescopes and large interferometers. Examine the visible light images of Mercury, Venus, the Earth, Mars, and the Moon, in Figure 13.9. Imagine the outcome of repeated measurement of the total sunlight (not the spectrum) reflected by the whole sunlit hemisphere of each of these planets. If these measurements are plotted versus time then we obtain what is called the light curve of the planet.

In the case of the Earth a typical light curve would show considerable variation over one rotation period (one day). Oceans reflect little light, clouds and surface ice reflect a lot, and deserts and vegetation reflect somewhere between. As the Earth rotates, the different contributions would be presented to a

(a)   (b)   (c)

(d)   (e)

FIGURE 13.9 Visible light images of (a) Mercury (NASA/JPL) (b) Venus (NASA/JPL) (c) the Earth (NASA/LPL) (d) Mars (HST/STScI) (e) the Moon (USGS).

distant observer in different proportions, to give a diurnal variation in the brightness of the illuminated hemisphere of up to a factor of two. The variation would be even greater if, instead of the total light reflected, light in certain bands of wavelength were isolated.

Light curves for the other planets (and the Moon) in Figure 13.9 display a smaller variation in reflected sunlight. This is because they have no oceans and either no cloud cover (Mercury and the Moon), or total cloud cover (Venus), or slight cloud cover most of the time (Mars).

There are longer term changes in the light curve when it is averaged over the planetary rotation period. In the case of the Earth these are up to 20%, due to day to day, week to week, and month to month variations in cloud cover. For the other bodies in Figure 13.9 these changes are smaller.

The light curve of an exoplanet will not give a very strong indication of whether it is habitable, or inhabited, but it would be indicative, and combined with other observations, such as the infrared spectrum, it could help build a convincing case for or against life being present.

Atmospheric absorption spectra at visible and near infrared wavelengths

As well as the reflection spectrum of the planet, we also have the *absorption* spectrum of its atmosphere at visible and near infrared wavelengths. This is the outcome of the light reflected by the planet's surface being absorbed at certain wavelengths by the atmospheric constituents. Like the infrared spectrum, the visible-infrared absorption spectrum can reveal the atmospheric constituents. O_2 can now be detected directly, because it has a strong spectral signature, particularly at the red 0.76 micrometer wavelength, unlike the weak spectral signature that it has in the infrared where we have to detect it by proxy through O_3. The feasibility of this approach has been demonstrated by examining the Earth's visible spectrum reflected off the dark side of the Moon (Earthshine), though to apply it to exoplanets we will again need giant optical telescopes.

As an aside, you might wonder why O_2 has a strong spectral signature at visible wavelengths, given that it is the symmetry of the molecule that prevents it having a strong infrared signature (Section 13.3). Recall that this is because in the infrared it is the atoms of oxygen as a whole that are influenced by the radiation, and so the symmetry is sensed. At the much shorter visible wavelengths a single electron in one of the atoms is involved and the molecular symmetry is not involved.

All investigations of an exoplanet are easier and will reveal more the closer we get. This requires that we send our instruments out of the Solar System. What are the prospects for interstellar journeys?

13.7 INTERSTELLAR PROBES

The obvious advantage of an interstellar probe (Figure 13.10) is that you can get much nearer to the exoplanet. Once there, it would be much easier and quicker to gather spectral data than from a remote Earth-based vantage point, and it also makes it possible to gather spectral information that would be too weak to get from Earth. If the probe entered the atmosphere and landed on the surface, then direct sampling and close up imaging would be possible, and any but the most deeply buried biospheres would be detected. So, as we have sent probes to the outer Solar System, why don't we send them to exoplanets?

The nearest star to us is Proxima Centauri, an M dwarf, part of a triple system that also includes the solar type star Alpha Centauri A and the slightly larger Alpha Centauri B. Proxima Centauri is 4.22 light years away, and slowly orbits the other two stars, which are much closer to each other than Proxima Centauri is to them. They are each 4.40 light years from us.

Consider sending a probe to Proxima Centauri. Though it is the closest star, it

FIGURE 13.10 Artwork showing an interstellar probe powered by nuclear propulsion. (NASA)

is about 267,000 times further than the Sun. At an average speed of 10% of the speed of light it would take 42.2 years for a probe to get there. This is rather a long journey! Moreover, the energy required to accelerate a probe to 10% of the speed of light is huge, and would need a propulsion method other than the chemical ones used predominantly today.

What have we achieved with space probes so far? Since our first steps into space in the 1950s, many probes have been launched, and though none of them has been aimed at the stars, there are five NASA spacecraft that will leave the Solar System, all but one of them having completed their missions to the outer planets. These are Pioneer 10, Pioneer 11, Voyager 1, Voyager 2, which explored the giant planets, and New Horizons, currently *en route* to the dwarf planet Pluto, which lies beyond Neptune, in an orbit with a semimajor axis of 39.8 AU. It will fly past Pluto in July 2015.

Though Pioneer 10 was the first of these spacecraft to be launched, in 1972, it is Voyager 1, launched in 1977, that is now furthest away. Since its flyby of Saturn in November 1980 it has been moving out of the orbital plane of the planets. In April 2007, it was 101.4 AU from the Sun, though this puts it only

0.16% of a light year from us! It has been slowed by the Sun's gravity to a speed of a few kilometers per second, and so it will be tens of thousands of years before it gets among the nearer stars, and very much longer before it accidentally gets close to one. Nevertheless, the possibility that it will be found by aliens has not been ignored. Voyager 1 carries a long playing gramophone record, bearing sounds and images of Earth. Whether any aliens could play and understand it is a matter of debate.

It is likely to be several decades before we find a feasible and affordable way to launch a probe that could achieve an average speed of 10% of the speed of light. There is also the problem of slowing the probe down so that it does not whizz past any planet in a few seconds. One possibility is to use radiation pressure from the solar photons acting on huge sails to accelerate the craft out of the Solar System, perhaps supplemented by a laser beam from Earth, and then slow the craft down with the photons from its destination star. Our endeavors would be facilitated by the use of tiny spacecraft that could easily fit in a thimble. Their very small mass would enable high speeds to be achieved for relatively small energy expenditure, provided that a suitable propulsion system could be developed that required little mass in itself to be carried well beyond the Solar System.

Even so, there is still the 42.2 year travel time to Proxima Centauri, plus the 4.2 years needed for information from the probe to reach us at the speed of light.

It is for a distant successor to this book to give interstellar probes more space. Fortunately, developments in spectroscopy from ground-based and space telescopes give us a real prospect of finding life beyond the Solar System in the next few decades by making observations from our home patch.

But we might not have to wait that long to discover that we are not alone. This discovery could even be made tomorrow, as you will see in the next chapter!

14

The search for extraterrestrial intelligence

There is one way in which we could discover extraterrestrial life now, with no more than the instruments we already have, and this is by life telling us that it is there. This could be by transmitting radio or laser signals that we receive, or by sending probes into the Solar System that we find, or in any other way that a technological intelligence could make its existence known. And indeed it is *technological* intelligence that we will have discovered. Regardless of the likelihood of such a discovery, it would be an astonishing finding. We would have learned not only that there is life in at least one other place in the Universe, not only that it had evolved into multicellular species, not only that one of these species had evolved intelligence, but also that the intelligence was technological. We would have discovered what is usually called extraterrestrial intelligence (ETI). Whatever they looked like, these creatures would be thinking like us – they would be trying to understand the Universe and manipulate their environment in the same way that we do. What are the chances of such a discovery, and how might it be made?

14.1 THE NUMBER OF TECHNOLOGICAL INTELLIGENCES IN THE GALAXY

The Galaxy contains about 200,000 million stars. There are thousands of millions of other galaxies in the observable Universe, of various types, some containing more stars, others fewer. Overall, the observable Universe contains such a huge number of stars that it is beyond reasonable doubt that it contains at least a few other technological intelligences. However, discovering any such intelligences beyond our own Galaxy is unlikely given the vast distances involved. It is therefore usual to focus the search for extraterrestrial intelligence (SETI, pronounced "setee") within our own Galaxy.

Consider the number N of detectable civilizations in the Galaxy today. There is an equation for N well known to astronomers. This is the Drake equation, named after the US astronomer and SETI pioneer Frank Drake who presented it at a meeting in the USA in 1961. Note that N is also the number of planets with detectable civilizations – it is assumed that a detectable civilization lives on a planet.

The Drake equation

The essence of the equation is that it displays the factors that determine the value of N. These are

- the rate R_{hp} at which habitable planets appear in the Galaxy;
- the fraction f_{bio} of these planets that have a biosphere;
- the fraction f_{ti} of biospheres in which species evolve that are technologically intelligent i.e., have the potential to be detected over interstellar distances; and
- the average time L_{det} for which a planet harboring technological species remains detectable.

I hope you can see that the larger the value of each factor the greater will be the value of N. Conversely, the smaller the value the lower the value of N.

R_{hp} is the astronomical factor. Estimates range down from about one per year to a few hundred times smaller. If that seems disconcertingly imprecise, the imprecision pales into insignificance alongside that in the remaining factors!

The factor f_{bio} is a biological factor. In principle, it could be as small as zero, corresponding to no habitable exoplanets with biospheres, or as large as 1, corresponding to every habitable planet having a biosphere. It is widely believed among astrobiologists that if a planet is habitable then it will probably have a biosphere, though our ignorance about the way life started on Earth makes some scientists dubious about this.

The factor f_{ti} is also biological. In principle, it too could be as small as zero, corresponding to no inhabited exoplanets having a technologically intelligent species, or as large as 1, corresponding to every inhabited planet having a technologically intelligent species. Unfortunately we have little idea of the value of this factor. Some scientists believe that if there is life then technological intelligence will almost certainly follow. Others believe that the emergence of technological intelligence is so unlikely that we could be the only such species in the Galaxy!

Things only get worse when we turn to the final factor, the average time L_{det} for which a planet harboring technological species remains detectable. We have been detectable through accidental leakages of radio, TV, and radar for a few decades. Deliberate attempts at interstellar communication have been confined to a handful of cases since the first attempt in 1974. Will we be detectable (through leakages or through deliberate attempts) for another 100, 10,000, 100,000 years, or far longer? We have no idea. Likewise we have no idea for how much longer our species will survive. *Homo Sapiens* has been around for about 100,000 years. But how much longer we will be here? Will it be a thousand, a million, or a thousand million years? Will there be other technological species to succeed us?

Box 14.1 The Drake equation

The material here is for readers with a basic knowledge of algebra. It is not essential.

The Drake equation gives a value for the number N of detectable civilizations in the Galaxy today. Here is the form that I am using

$$N = R_{hp}\, f_{bio}\, f_{ti}\, L_{det}$$

These symbols have been defined and discussed in the main text. You can see that to obtain a value for N the values for the factors have to be multiplied together. It is implicit in the Drake equation that none of these factors has varied since the Galaxy was born. This is not the case, but in considering to what extent, if at all, we can use the equation to estimate N, this assumption will suffice.

To obtain insight into this equation here it is in another form

$$N = R_{ti}\, L_{det}$$

where $R_{ti} = R_{hp}\, f_{bio}\, f_{ti}$, which is thus the (assumed constant) rate at which planets harboring technological species appear in the Galaxy. R_{ti} is in units of the number of technological species appearing per year (or per any other unit of time).

The Figure shows that, starting from zero, technological species appear at the rate R_{ti}. In the time L_{det} the number that have appeared is $R_{ti}\, L_{det}$. Thereafter, for each new appearance, on average another species comes to the end of its detectable lifetime L_{det}, and so $R_{ti}\, L_{det}$ is the steady state value of N.

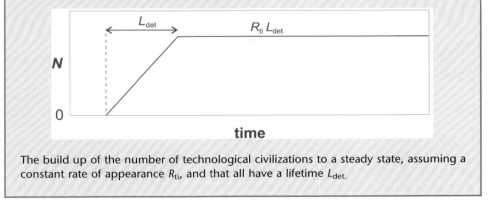

The build up of the number of technological civilizations to a steady state, assuming a constant rate of appearance R_{ti}, and that all have a lifetime L_{det}.

The huge uncertainty in L_{det}, coupled with that in f_{ti}, means that the number N of detectable civilizations in the Galaxy could be anything from 1 (which is us alone) to hundreds of millions. Therefore, the Drake equation is not useful for calculating N. Its use is that it displays the important factors. The safe conclusion is that SETI is, in essence, an experimental science – if we want to know we have to search.

14.2   SEARCHING FOR EXTRATERRESTRIAL INTELLIGENCE (SETI)

There are three ways in which we could detect extraterrestrial intelligence. We could

- detect radiation that travels from them to us;
- detect their spacecraft or other artefacts in the Solar System (even the aliens themselves!); or
- detect features of their planet, or of their environment in general, that reveal technological modification.

In considering each of these I shall often use our own technological capability as a point of reference. We have only recently become detectable across interstellar distances, and our ability to communicate with the stars is slight. It is likely, given that the age of the Galaxy is about three times the age of the Solar System, that another intelligence will be older and therefore more advanced than us, so if a means of communication looks feasible to us it is likely to be feasible to them, and very probably something they could implement easily.

What sort of radiation?

Though there are other types of radiation that travel across space (such as gravitational waves and cosmic rays), electromagnetic radiation seems best for interstellar communication. This is because

- it is relatively easy to generate, to beam at a selected target, and to detect;
- there are wavelength ranges that are only weakly absorbed or scattered by the interstellar medium, and only weakly absorbed by planetary atmospheres; and
- information can readily be imprinted on the radiation, and with little energy being required to send each item of information.

The one problem with electromagnetic radiation is that it takes a long time for messages to travel over interstellar distances. It travels at the speed of light (denoted by c), and though by terrestrial standards this is an enormous 299,792 km per second, it still takes 4.22 years to reach us from the nearest star. The current laws of physics do not rule out faster-than-light travel (superluminal travel), so could we use speedier means?

One possibility is particles called tachyons. These could only travel faster than light. Just as ordinary matter cannot be accelerated to speeds as great as c, tachyons cannot be slowed down to c. Unfortunately they remain only a theoretical possibility, as yet unobserved. Another possibility relies on a phenomenon called quantum entanglement. In quantum theory it seems possible that two particles can be inextricably linked, so that if something happens to one, then something happens to the other instantaneously, no matter how far away it is. If this phenomenon could be utilized then there could be instantaneous communication, though for the moment it is just another

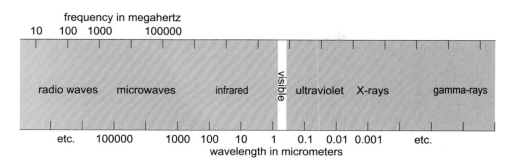

FIGURE 14.1 The electromagnetic spectrum. (The same as Figure 2.9, except that the spectrum has been reversed. This is to facilitate comparison with Figure 14.3, which is usually plotted with frequency increasing to the right. Also, the wave frequencies have been added.)

theoretical possibility. For practical purposes, electromagnetic waves offer the fastest means of communication that we know of at present.

Figure 14.1 shows the electromagnetic spectrum as a strip diagram. It covers a huge range of wavelengths (or frequencies – a megahertz is one million cycles per second). The question arises, which parts of this spectrum are best for interstellar communication. In the late 1950s, which mark the start of the modern era of the search for extraterrestrial intelligence, SETI, the answer was "microwaves and optical", and this remains the answer today, though for the first few decades it was microwave searches that dominated SETI. It was only in the mid-1990s that searches began to proliferate in the optical part of the spectrum, particularly at visible wavelengths (the optical range also covers the infrared and the ultraviolet). Consider these searches, first, at microwavelengths.

14.3 SEARCHES AT MICROWAVELENGTHS

Figure 14.2 is a spectrum of the power received from the cosmos over a range of frequencies that encompasses microwaves. It illustrates some of the reasons why microwaves have been favored for SETI. Three distinct components of the overall spectrum are shown. One shows the power in the radiation that we typically receive from the Galaxy. This is due to what is called synchrotron radiation from electrons as they move through the magnetic fields that pervade our Galaxy. It declines steeply as frequency rises, until it is exceeded by microwave radiation left over from the Big Bang. At even higher frequencies, absorption in the Earth's atmosphere raises the power threshold for the detection of extraterrestrial sources. The solid curve shows these three components combined. In addition, there is an increase in microwave receiver noise as frequency increases, and this exceeds the effect of the Big Bang radiation at frequencies greater than about 10 000 megahertz, though it does not rival atmospheric absorption.

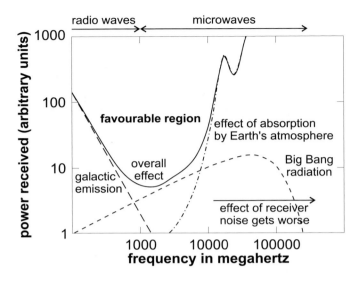

FIGURE 14.2 The microwave spectrum, and adjacent parts of the infrared and radio ranges of the electromagnetic spectrum, showing the microwave 'window' for SETI, labeled as the favorable region.

A favorable region emerges that spans the lower microwave frequencies and the highest radio frequencies, in what is called the microwave window. The region around 1000 megahertz is best. Here the background radiation is at its lowest (except for terrestrial broadcasts in some frequency ranges), and absorption in the Earth's atmosphere is very slight. In addition, the photon energy, which is proportional to the frequency, is comparatively low at such low frequencies. Variations with time in the photon flux could be used to convey information. Therefore, messages can be sent by microwaves with low energy per item of information.

Consequently, the very first SETI in the modern era was made at these microwave frequencies. This was Project Ozma in 1960, led by Frank Drake, and carried out with a radiotelescope at Green Bank, West Virginia, with a 25 meter diameter dish (Figure 14.3). A total time of 150 hours was spent looking at the nearby stars Tau Ceti (a G dwarf, like the Sun) and Epsilon Eridani (a K dwarf) at frequencies around 1420 megahertz. No signals from ETI were found.

The choice of 1420 megahertz was not arbitrary. But why make a choice at all? Why not have a detector that spans all of the frequencies in the microwave window at the same time? The reason is illustrated in Figure 14.4, which shows the advantage of ETI confining its transmission to a narrow range of frequencies. This helps the transmission to outshine the natural background. If neither background nor signal fluctuated randomly this would be less of a requirement. But they do fluctuate. The range of frequencies that the signal covers is called its bandwidth. Likewise, the range that the receiver detects is also called its

FIGURE 14.3 The Project Ozma team at a reunion some years ago, under the Green Bank 25 meter telescope with which they made the first modern search for ETI. Frank Drake is second from the right in the back row. (Image courtesy of the National Radio Astronomy Observatory/AUI/NSF)

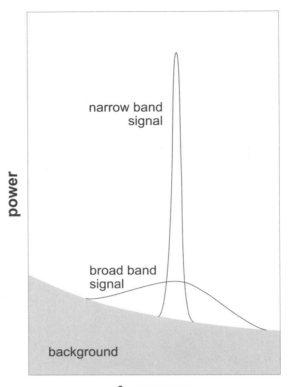

frequency

FIGURE 14.4 The advantage of a narrow bandwidth transmission in outshining the background.

bandwidth. In Project Ozma the receiver bandwidth was 100 Hz, which is the tiny fraction 1/100,000,000 of the microwave window. Even so, you might wonder if it would have been possible to scan the receiver across the window. This was not practical, if only because of the enormous time it would have taken to search 100,000,000 frequency bands. Consequently, the receiver was scanned only in the vicinity of 1420 megahertz.

This region was chosen because radiation at 1420.4 megahertz is emitted by a hydrogen atom when it undergoes a transition in which the spins of the electron and the nucleus (a proton) go from having their rotation axes parallel (pointing in the same direction) to antiparallel (pointing in opposite directions). The rationale was that this spectral line is one of the most clear and widespread emitted by the interstellar medium, thus providing ETI a basis for selection from among the host of other microwave frequencies. Communication would be more likely *near* 1420.4 megahertz rather than at it, to facilitate discrimination from the natural sources of the line.

Recent and current searches

Since Project Ozma there have been a few hundred microwave searches, some of them lasting many years. Many have used radiotelescopes larger than the 25 meter Green Bank telescope – a larger dish can detect smaller signals i.e., the sensitivity is greater (as with optical telescopes, Section 8.1). Bandwidths less than 1 hertz have been used in some cases, coupled with receiving systems that can examine tens or hundreds of millions of adjacent bandwidths (channels) simultaneously.

One such was the Billion Channel Extraterrestrial Assay (BETA), master-minded by the US physicist Paul Horowitz, which used the 26 meter radiotelescope at Oak Ridge USA from 1995 until the radiotelescope was blown over by strong winds in March 1999. BETA used a bandwidth of 0.5 hertz over the range 1400–1720 megahertz i.e., 640 million channels – not quite a billion (1000 million). Also in 1995 the privately funded SETI Institute in California launched Project Phoenix, which had risen from the ashes of the canceled NASA search Microwave Observing Program (MOP). Between 1995 and the end of the project in March 2004, among other radiotelescopes, it used small proportions of the observing time of the largest single dish radiotelescope in the world, the 305 meter dish at Arecibo in Puerto Rico, and the 76 meter Lovell Radiotelescope at Jodrell Bank in the UK (Figures 14.5 and 14.6). Project Phoenix was the world's most sensitive and comprehensive search for extraterrestrial intelligence, which scrutinized about 800 nearby stars not very different from the Sun, over the frequency range 1200–3000 megahertz.

SERENDIP IV (Search for Extraterrestrial Radio Emissions from Nearby Developed Independent Populations) is a project in which anyone with a computer connected to the internet can take part. A detection system is permanently connected to the Arecibo radiotelescope. Without interfering appreciably with other observations, it taps the 1420 megahertz signal in channels 0.6 hertz wide, from 1.25 megahertz below 1420 megahertz to 1.25 megahertz above it, from whatever part of the sky the telescope is looking at while devoted to other projects. A screensaver called SETI@home takes a small portion of the data, analyzes it, and returns the analysis to the main SERENDIP computer. Details can be obtained via the website listed in Further Reading and Other Resources.

The maximum distance to ETI at which we could detect a signal, depends on

- the power they feed into the transmitted signal;
- the bandwidth of their signal, the narrower the better (Figure 14.4);
- the angular spread of the beam from their radiotelescope, the smaller the better, so that the power is more concentrated on their target – the spread decreases as the dish diameter increases, so, the larger the telescope the better;
- the diameter of *our* radiotelescope, the larger the better because we collect more radiation in a given time (as with optical telescopes – Section 8.1); and
- the bandwidth of our radiotelescope receiver, the closer to the bandwidth of the signal the better, so that as little extraneous power as possible is collected that would only dilute the signal.

FIGURE 14.5 The 305 meter dish of the Arecibo radiotelescope in Puerto Rico. (Photo courtesy of the NAIC-Arecibo Observatory, a facility of the NSF. Photo by Tony Acevedo)

The crucial point is that we have the capability to detect signals from twins of our larger radiotelescopes, across a substantial fraction of the diameter of the Galaxy, the disc of which is about 100,000 light years across (Section 7.1). ETI could easily have radiotelescopes far bigger than our own, in which case signals from ETI could be detected by us, now, from anywhere in our Galaxy of 200 thousand million stars.

What a microwave signal from ETI might be like

A signal is sought that is clearly not of natural or terrestrial origin. A terrestrial origin can be ruled out if the signal comes from a particular location among the stars. To see whether this is the case requires either that the signal is observed with at least two widely separated telescopes, or that the signal lasts long enough for the Earth's rotation to change the direction to a source among the stars – a few minutes is plenty long enough.

To rule out a natural origin we need to be aware of pulsars (Section 11.1). These constitute a sub-class of neutron stars, which are remnants of massive

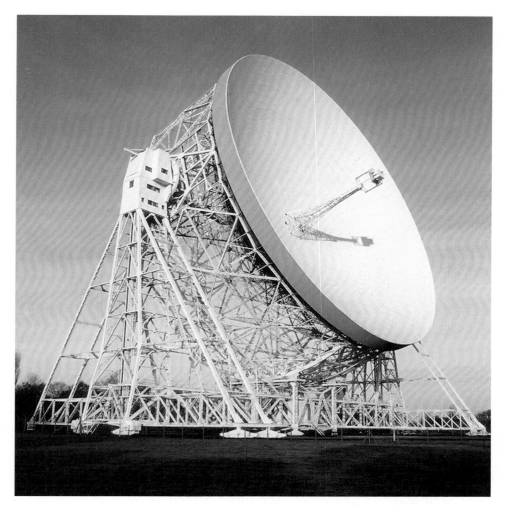

FIGURE 14.6 The 76 meter Lovell Radiotelescope at Jodrell Bank, UK. Part of its observing time is devoted to the SETI Project Phoenix. (Jodrell Bank Observatory)

stars. A neutron star can emit a beam of microwaves (and other wavelengths) at an angle to its rotation axis, as illustrated in Figure 14.7(a). If, as the star rotates, the beam sweeps across the Solar System, then we can observe a regular series of pulses as in Figure 14.7(b), and the neutron star is seen as a pulsar. Each pulse consists of a short train of microwaves. Pulsar rotation periods range from a few seconds down to about a millisecond, so this is the range of pulse spacings. If we did not know about rotating neutron stars then we might fall into the trap of thinking we had detected a transmission from ETI. Indeed, when the first pulsar was discovered in 1967 this possibility was considered, though only briefly. The explanation in terms of a rotating neutron star was soon at hand.

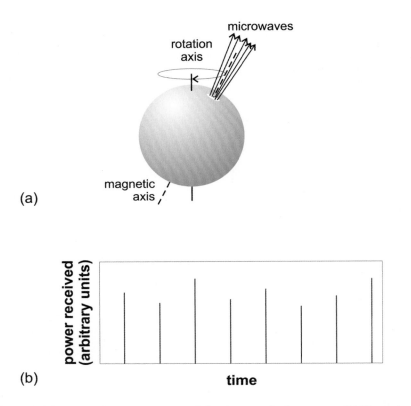

FIGURE 14.7 (a) A rotating neutron star emitting beams of microwaves. (b) The train of pulses that could be observed on Earth. Note that the pulse height varies randomly.

A telltale absence from a pulsar signal is *information* – the pulses carry no message. They could do so if the spacing of the pulses varied in accord with some code, but they beat as constantly as a highly accurate clock. One way to imprint a message on a signal is in a binary code. This is any code made up of two symbols, one of which can be regarded as 0 (zero), the other as 1. For example, the frequency of microwaves could be switched between two adjacent values, as in Figure 14.8, in which case one frequency could represent 0 and the other 1. The great advantages of a binary code are that it is readily recognisable, and that the imprint on microwaves is comparatively well preserved across interstellar distances.

Though some natural process might generate microwaves that alternate between two frequencies, it would not generate a meaningful sequence of the binary digits 0 and 1. Our first guess might be that they constituted a string of numbers rather than symbols in some alien script. Thus, in decimal terms, 0 and 1 have their usual meanings, but 10 is 2, 11 is 3, 100 is 4, and so on (or possibly 01 and 001 for 2 and 4 if we are meant to read from right to left). Proceeding on this assumption, suppose that the string could then be interpreted (in decimal

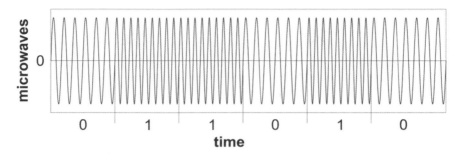

FIGURE 14.8 Microwaves coded in binary through the frequency of the wave varying between two adjacent values, and the corresponding sequence of zeroes and ones. The frequency difference is exaggerated here, and there would be far more cycles between frequency changes.

form) as the digits 31415927 repeated several times. These are the first 8 digits in the ratio of the circumference of a circle to its diameter, 3.1415927 (this ratio is given the Greek symbol π, "pie"). It would then be beyond reasonable doubt that we had detected a signal from ETI (in this example, "pie in the sky"!).

To summarize, we expect a microwave signal from ETI to be narrow band, and to be coded in a comparatively recognisable way (probably binary) with a message that would be distinct from the imprint of any natural process (e.g., the digits of π). Whether we could understand any content beyond universal mathematics is a matter of debate, but remember that if ETI is trying to communicate then it will make the message as easily decodable as possible. Quite what we might learn, and what the effect would be on us, is unknown, but would surely be profound. First, however, we must detect a signal from ETI.

The outcome of microwave searches, and their future

Though some peculiar signals have been observed, all have been brief and unrepeated, and none has borne the imprint of an artificial origin. Therefore there is no evidence for ETI in the microwave searches to date. One explanation is that we are alone. Another is that ETI transmissions are too weak or too broadband. This is unlikely to be the case for deliberate attempts at communication, but probably would be the case if the transmissions were intended only to be detectable within their planetary system or between neighboring systems.

This latter case puts us in the eavesdropping mode of detection. It has been possible for ETI to eavesdrop on us ever since we began to make radio and television broadcasts, and to use radar. Ignoring the early, very weak transmissions, a shell of microwaves and radiowaves has traveled nearly 100 light years from the Earth, encompassing nearly 10 000 stars. These transmissions are however still comparatively weak, and as they spread out they become even weaker. Moreover, the strongest of our transmissions, radar, has bandwidths of about 1000 hertz.

Consequently it is difficult to pick out the signal from beyond the Solar System. Even from the nearest stars, ETI would have to have receivers far more sensitive that ours to detect our leakages, and the comparatively large bandwidths we use might cause ETI to fail to recognize them as artificial. It is also possible that the free air broadcast and communication channels that we use are a brief phase. Indeed, an increasing number of broadcasts and communications are now carried along wires or optical fibres, and therefore leak nothing to space.

But if ETI is out there and is trying to communicate, then the likely reasons for failure include the limited number of narrow frequency bands that we have searched, and the limited number of stars that we have so far targeted. There is also the limited time for which we have targeted each star, assuming that ETI sweeps a narrow beam around the sky, so that the beam dwells on us for only a small proportion of the time.

Microwave searches continue, and there are plans for much larger ones. Notable is the Allen Telescope Array (ATA) under construction at Hat Creek Observatory, in northern California. It is funded by the SETI Institute in California, and will be dedicated to SETI. There will be at least 350 dishes, each 6.1 meters in diameter, which will give it a large collecting area of about 10,000 square meters. This is only about a seventh that of the Arecibo telescope, but the ATA can simultaneously cover 16 targets over the range 500–11,000 megahertz. By February 2007, 42 dishes had been completed and are now making observations. The ATA might be fully operational in 2008, though this depends on funding. A smaller array is being constructed by the SETI League in New Jersey

FIGURE 14.9 The Allen Telescope Array, as it will look on completion.

USA. If these and other radiotelescopes fail to discover ETI by, say, 2025, then we could conclude either that ETI is extremely rare in the Galaxy, or that ETI rarely attempts microwave interstellar communication.

14.4   SEARCHES AT OPTICAL WAVELENGTHS (OSETI)

Recall that optical wavelengths span not only visible radiation, but also ultraviolet and infrared radiation (Figure 14.1). However, in optical SETI (OSETI) it is visible radiation that has so far dominated the searches, and the following discussion is therefore restricted to *visible* OSETI.

At first sight visible wavelengths do not seem promising. Though the Earth's atmosphere is transparent at visible wavelengths, there are copious natural sources in the form of stars and glowing interstellar gas. Moreover, any planets around all but the nearer stars would have a very small angular separation from their star, beyond our present capability to obtain images sharp enough to allow us to distinguish light from the planet from light from the star. Consequently it might seem that optical signals from the planet would be overwhelmed by the light from the star.

These were the beliefs of many scientists until the 1990s, which is why microwaves had been preferred up to that time. The device that makes OSETI worthwhile is the high power laser. Laser light has many remarkable properties, but the ones of most relevance here are that:

- extremely tightly collimated beams can be produced, with angular diameters considerably less than a second of arc (3,600 seconds of arc = 1°); and
- extremely high powers can be achieved.

By 1996 powers had reached just over a petawatt within well-separated pulses each lasting a few tenths of a picosecond (1 petawatt = 1,000,000 gigawatts, where 1 gigawatt = 1,000,000,000 watts, and 1 picosecond = 1/1000,000,000,000 seconds). Within the angular spread of the beam, and during the pulse, this exceeds the brightness of a beam with a similar angular spread from a solar-type star over its whole spectrum! Doubtless, ETI could build yet more powerful lasers.

OSETI signals

Figure 14.10 shows what a laser beam leaving a planet might look like. Figure 14.11 shows the way the information could be carried within it. Suppose that the huge pulses have peak power of 1,000 petawatts, but are only 1,000 picoseconds in duration and are spaced by 1 second. The space between the pulses is thus 1,000,000,000 times longer. Thus the *average* power over each second is 1,000 petawatts/1000,000,000. This is 1/1,000,000 petawatts, or 1 gigawatt. This is about the output of a large power station, so could be regarded as well within the capacity of ETI not much in advance of us. More rapid pulsing, and thus more rapid communication, would require greater average power.

FIGURE 14.10 A laser beam leaving a planet.

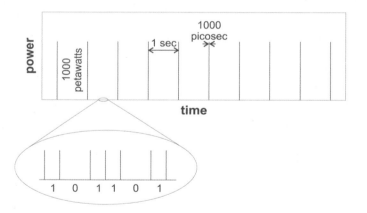

FIGURE 14.11 A possible form of information carried within the beam, by pulses. Note that each giant 1,000 picosecond pulse would contain about half a million wavelengths of visible light. It is the weaker pulses (between the giant pulses) that deliver a message.

If the beam width of the giant pulses was around 0.01 seconds of arc (which is feasible), then a 10 meter optical telescope on Earth would easily detect these pulses at a range up to a few hundred light years. Moreover, there is no problem of which frequency to select – even though each 1,000 picosecond pulse contains about half a million visible wavelengths, and thus has a well-defined frequency, the signal, even at a few hundred light years, would be detectable without filtering out other frequencies.

These giant pulses, if regularly spaced, would, like those from pulsars, carry no message. Even if the spacing was varied, the message would arrive very slowly. The huge pulses do, however, grab our attention, and ETI could have included weaker pulses that arrive at a faster rate (Figure 14.11). We would now notice these pulses, and variations in their spacing would deliver a message quickly. The weaker message pulses would benefit from frequency filtering, though this could be broadband.

Within about 400 light years of the Sun there are roughly half a million stars. Not all of these will be accessible to OSETI, because, unlike microwaves, visible radiation is absorbed and scattered by dust. Therefore, dusty regions will be inaccessible, and this is an important restriction at greater ranges, where there are dusty regions such as the Orion Nebula (Figure 14.12) and the central regions of the Galaxy.

FIGURE 14.12 The Orion Nebula, a huge interstellar cloud of gas and dust from which stars are forming. It is about 1,500 light years away and about 15 light years across. (NASA/ESA)

Another consideration for OSETI arises from the very narrowness of the laser beams. If ETI is targeting a particular star, they have to know where the star will be when their optical signal reaches it. With 0.01 seconds of arc collimation, the beam is only about 1.2 AU across even at a range of 400 light years. Therefore, ETI at this distance, will have to know where the target star will be in 400 years, and they have to know it to an accuracy of about 1.2 AU. Only in recent years have *we* acquired sufficiently accurate data on stars.

Searches now and in the future

Only a small amount of effort has been devoted to OSETI, but in recent years the activity has been growing, in part following the development of high power pulsed lasers.

An early pioneer is the British engineer Stuart Kingsley, who has operated the Columbus OSETI Observatory in the USA since the early 1990s, searching for laser pulses, so far without success. It is privately owned and run, and based on a 0.25 meter optical telescope.

A Harvard–Smithsonian group under Paul Horowitz performed OSETI from October 1998 to November 1999 by piggybacking on Harvard's 1.55 meter telescope at Oak Ridge in the USA, whilst it surveyed about 2,500 stars. Horowitz also searched (unsuccessfully) for laser pulses. Other ongoing searches include one at the University of California at Berkeley, using a 0.76 meter telescope, again to detect laser pulses. This is under the direction of the US astronomer Dan Werthimer.

A different type of search is in progress at Berkeley, and involves examination of records of optical spectra taken during searches for extrasolar planets. Under the direction of US astronomer and exoplanet hunter Geoffrey Marcy, the records are being scrutinized for a *continuous* laser signal, rather than a pulsed laser signal.

The number of optical searches is growing, and several more are planned for the near future. For example, the Harvard–Smithsonian group has combined with a group from Princeton University to mount a detector of laser pulses on Princeton's 0.91 meter telescope. The Harvard and Princeton telescopes will scrutinize targets at the same time, to help avoid spurious results. Additionally, the Harvard–Smithsonian group is building a system based on a new, 1.8 meter telescope at Oak Ridge, to be dedicated to OSETI – this will also search for laser pulses. In addition, it is to be hoped that at least a small proportion of the time on the world's largest telescopes will be devoted to OSETI. This will increase the chance of the detection of ETI, provided always that they are out there trying to make contact.

14.5   SPACECRAFT, AND OTHER ARTEFACTS FROM ETI

In spite of many claims in the popular media, there is no evidence that the Solar System has ever been visited by ETI, either in person or robotically. If, as seems likely, there have indeed been no visitations, is this because interstellar travel is extremely difficult? Is it so difficult that we are unlikely ever to find ETI in this way? To answer this question, consider how difficult it would be for *us* to travel to the stars.

Interstellar travel

The fundamental requirement is for the spacecraft to acquire sufficient speed for it to reach the stars in a reasonable time. Consider first the energy needed to combat gravity.

To get away from the Earth to interstellar space, most of the energy needed is to combat the Sun's gravity. For a payload of just one kg the energy amounts to about 1,000 million joules. A joule is a unit of energy. To get a feel for the size of the unit, note that in lifting a heavy person of 100 kg from the Earth's surface to a height of 100 meters requires just under 0.1 million joules, so you can see that a *lot* of energy is required to escape the Solar System. Moreover, to the mass of the payload we must add the mass of the chemical rocket fuel that accompanies the payload, which initially accounts for nearly all of the mass of the spacecraft.

If we give the spacecraft just enough energy to reach interstellar space it will arrive there traveling very slowly. We must do better than this. About the best we could hope to do at present would be to deliver the spacecraft with a speed of about 100 km per second, which is only 0.033% of the speed of light. Even for this modest speed, an additional energy of 5,000 million joules per kg is required. This makes 6,000 million joules to deliver a meagre 1 kg of payload to interstellar space, traveling at 100 km per second. To this must be added the energy needed to carry the chemical fuel, which will be several times 6000 million joules. Also remember that sufficient fuel must be carried to the destination to slow the payload down.

At 100 km per second it would take about 12,600 years to reach the nearest star, Proxima Centauri. At 10% of the speed of light the time reduces to 42.2 years, but the energy per kilogram has risen to 450 million million joules! This means that, to deliver 1 kg to interstellar space at 10% of the speed of light, and adding the energy needed to accelerate the fuel, the energy requirement is the equivalent of the whole energy output of a large electricity power station for a couple of years.

The energy required from the fuel can be reduced if some other source of energy can be tapped. One such source is gravity boosts involving the Sun and Jupiter. In a gravity boost the spacecraft encounters a body in such a way that the kinetic energy of the spacecraft is increased at the expense of the kinetic energy of the body. The mass of Jupiter and the Sun is so great that their change in orbital speed is negligible. These boosts reduce significantly the mass of chemical fuel that needs to be carried by the spacecraft, though a prodigious amount is still required.

To make a practical approach to 10% of the speed of light we need to substitute chemical fuels with something far more potent, delivering far more energy per kg of fuel. One possibility is nuclear fusion involving the hydrogen isotope deuterium (^{2}H) and either tritium (^{3}H) or the very scarce helium isotope ^{3}He. Per kg of these isotopes, about 10 million times more energy is released than in chemical reactions, and this release need not be explosive.

Even more energy, about 200 times more, is released from the annihilation of matter with antimatter. Antimatter differs from matter in that certain of its properties are reversed. As in the case of matter it has positive mass, but, for example, opposite electric charge, so that the anti-proton is negatively charged, the anti-electron (positron) is positively charged, and so on. When a particle meets its antiparticle both are annihilated and their total masses are converted into energy, in accord with Einstein's famous equation $E = mc^2$ where c is the speed of light. This gives an astonishing 90,000 million million joules per kg. Only about 0.005 kg of antimatter would be needed to accelerate one kg to 10% of the speed of light.

Antimatter is rare in the Solar System, and probably throughout the observable Universe. It appears in various nuclear reactions, but it is difficult to store – if it meets ordinary matter it is, of course, annihilated.

A completely different approach is solar sailing (Figure 14.13), in which the spacecraft is pushed along by the pressure on a huge sail exerted by the photons from a powerful source. Within the Solar System the Sun's radiation would be the source of photons, but in interstellar space it would be better to blow the craft along with a laser beam from the Earth. This could still be regarded as solar sailing if the laser was solar powered, such as by photovoltaic cells.

Exotic possibilities for interstellar travel

Wormholes are short cuts through spacetime, which is the four dimensional construct in relativity consisting of the three space dimensions and time. A wormhole is a channel that permits travel from one point in spacetime to another without us having to travel through ordinary space. This permits superluminal (faster than light) travel. It also permits travel back in time, which would lead to all sorts of problems – any slight alteration in history made by a time traveler could alter the world they traveled from, including the possibility that they did not now exist and so could not have traveled back in time to make the alteration! Therefore, though general relativity permits the existence of wormholes, there would surely have to be some barrier to travel backwards in time. Wormholes would be unstable, so to stabilize them they would need to be threaded by matter with the curious property of negative energy density. Quantum physics permits the existence of such matter, but so far, like wormholes themselves, it has not been found.

Another possibility is warp drive, in which a spacecraft warps spacetime in its vicinity in such a way that the spacecraft moves superluminally from one point in spacetime to another. As with wormholes, backwards time travel would need to be ruled out, and matter with negative energy density would be needed to maintain stability. According to *Star Trek*, warp drive lies a few centuries off. In

FIGURE 14.13 Solar sailing to the stars. (Batakin Space Center, courtesy of The Planetary Society)

fact, we cannot be sure that it is feasible at all. On the other hand, we know that our physics is incomplete, so we cannot be certain about future possibilities.

In conclusion, with our present technology, we are far short of being able to travel to the nearest stars with reasonable journey times on a human time scale. You might therefore be surprised to learn that we could embark on accomplishing the far grander exploration of the whole Galaxy. If newcomers to space technology like us are on this threshold, then maybe ETI is already well advanced in this enterprise. How could we embark on Galactic exploration? What are the consequences of ETI's greater capabilities in this respect?

Galactic exploration

In the 1940s and 1950s the Hungarian-born US mathematician John von Neumann developed the idea of the self-replicating probe. This has the capability of making a copy of itself from materials in its environment. As applied to galactic exploration any such probe sent from Earth to a nearby star would, on arrival, make more than one copy of itself from local materials. These copies would then be launched toward other stars. Meanwhile, the original probe would have transmitted information to Earth about its own destination star and any planets. When the second generation probes arrived at their stars the process would be repeated. This exponential growth in the number of probes would enable the whole Galaxy to be explored in roughly 100 million years even with our present chemical rocket technology. This is only about 1% of the age of the Galaxy. The one thing we lack at present is, understandably, the will to devote the huge resources needed to make such a probe.

More fancifully, a probe could also colonize a planetary system. It could carry the instructions to make human beings, and it could make a life supporting environment for them to inhabit. These would take the form of huge space stations, thus creating O'Neill colonies, named after the US physicist Gerard K O'Neill, who showed in 1974 how such large space colonies could be built.

If we could complete our exploration of the Galaxy by a time from now that is only 1% of its age, then any ETI that started exploration a relatively short time in the past could have largely or fully completed the exploration already. If so, then there might already be ETI artefacts in the Solar System. Some scientists have speculated that the asteroid belt would be a good place to look. Are all of those tiny asteroids really just lumps of rock?

Overall, it might not be worth making a huge effort to find alien artefacts, but we should at least be aware of the possibility of accidental discovery. What about detecting the effects of ETI's technology on their own environment?

14.6 TECHNOLOGICAL MODIFICATIONS BY ETI OF THEIR ENVIRONMENT

In 1964 the Russian astronomer Nikolai Kardashev classified civilizations into three types, based on their energy source.

- Type I civilizations utilize the energy resources of their planet, including the stellar radiation it receives. We are a Type I civilization.
- Type II utilize a large proportion of the whole energy output of the host star, not just the tiny fraction that falls on their planet.
- Type III utilize a large proportion of the energy output of their galaxy.

Each type could be detected from afar through its technological modification of its environment. Type I would be hard to detect. You have already seen in Section 14.3 that eavesdropping on communications is unlikely to be successful, with defense radar in our own case offering the best prospect to ETI. Another possibility is the gamma radiation from nuclear fission or fusion, though if we could detect such radiation then it would be likely to be at deadly levels for ETI, perhaps as a result of nuclear conflict – a depressing thought.

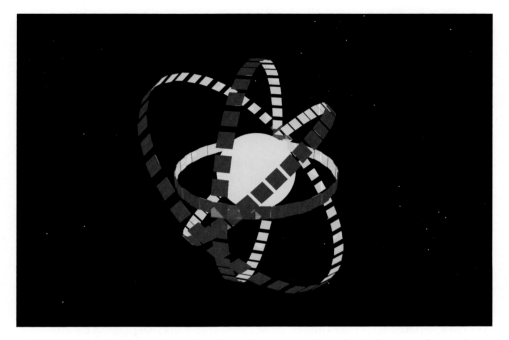

FIGURE 14.14 A Dyson sphere, diverting a large proportion of a star's energy for use by a Type II civilization. (Anders Sandberg, Oxford University)

Type II civilizations are more detectable. In 1959 the British physicist Freeman J Dyson envisaged ETI surrounding its star with a swarm of artefacts that divert a large proportion of the star's energy for use (Figure 14.14). A Dyson sphere (which is *not* a solid, complete sphere) utilizes the stellar radiation it intercepts. The energy flows through various energy conversion devices, to produce electricity, fuel, and so on. It then emerges ultimately as waste heat, just as for any energy conversion device. In equilibrium, no energy accumulates in the sphere. Instead, some fraction of the star's radiation, which is predominantly at visible and near infrared wavelengths, will be utilized and be radiated to space by the warmed sphere at middle and far infrared wavelengths. The star would seem to have a huge infrared excess. Such stars are known, but there is always a natural explanation, such as a star surrounded by a dusty envelope that absorbs the stellar radiation, that warms the dust, which then radiates at infrared wavelengths in accord with its temperature.

A Type III civilization might be detected through its reorganization of a galaxy! But though there is quite a menagerie of galaxies, none as yet requires such an extraordinary explanation. The galaxy wide communications that a Type III civilization presumably needs, might be powerful enough to be detectable in eavesdropping mode, unless electromagnetic radiation or any other type of radiation that we could detect, was no longer the basis of their communications.

14.7   THE FERMI PARADOX

You have seen that as yet there is no observational evidence for ETI. This does not rule out the existence of ETI. We might have failed to search for microwave or optical transmissions at the right time, or from the right star, or at the right frequencies. We might have failed to find artefacts in the Solar System, even though they are there. We might have failed to detect Dyson spheres, even though they are there too. Others argue that the failure to detect ETI is simply because ETI does not exist!

Around 1950 the Italian physicist Enrico Fermi posed this failure as the paradox that now carries his name. Fermi's paradox states that if ETI exists, then it must be widespread, in which case why aren't they among us? The assumption is that there has been plenty of time for ETI to have emerged and to have colonized the Galaxy – you saw in Section 14.5 that this would take a very small fraction of the age of the Galaxy. An obvious solution to the paradox is that there are no other technological species in the Galaxy.

Many and varied are the refutations of this bleak conclusion. One extraordinary hypothesis is that ETI is among us, looking human! This is difficult to test, difficult to disprove. As far as I know, no alien biochemistry with a human body has ever been found, and it is my belief that it never will be. Another hypothesis is that Galaxy wide ETI will not make its existence known to us until we are sufficiently advanced to join the galactic community. Frank Drake offers the less provocative hypothesis that it makes far better sense for technological species to colonize their own planetary system than endure the costs and hazards of going to other stars. Another possibility is that ETI is strongly disinclined to launch self-replicating probes, because the software controlling replication would certainly develop faults, and who knows what rogue behavior could result?

It is also possible that ETI has only just emerged in the Galaxy, and that we are among the first technological civilizations. This puts us in a rather special place in galactic history, but it cannot be ruled out. Even if we are not among the very first, ETI could still be thinly dispersed, particularly if the imperative to explore and colonize is weak. There might even be little desire to communicate. On Earth there is considerable reluctance to signal our presence across interstellar space, for fear of the consequences. Perhaps, therefore, our first contact will be with a doomed civilization that has little to lose by making its presence known.

There are many more refutations to the conclusion that we are the only technologically intelligent civilization in the Galaxy, not the least of which is that we have not yet searched hard enough or long enough. Certainly, if ETI is trying to make its communications obvious even to recently emerged species like ours, then we have some hope of success.

SETI must therefore be treated as an observational quest. If we want to find, we must search, though how much scientific effort should be devoted to the search is another question.

The consequences of success

What will happen were we to make an unambiguous detection of ETI? Aside from visitation by ETI, which is unlikely, this discovery will be made within the astronomical community. The lines of communication within this community are many and rapid. Therefore, governments would not be able to keep the discovery from their general public, even supposing that any wants to.

A discovery would certainly alter our sense of humanity's place in the Universe. Quite in what ways I can only speculate. Doubtless some would be uneasy about the presence of ETI, probably more advanced than us, even if it was far off. Others would be glad to know that we are not the only technologically intelligent species in the Galaxy. Perhaps we would learn from ETI things that are of huge scientific or practical importance, good and bad. Or perhaps we would be bewildered, or unable to understand what they are telling us. The discovery could even help humans to live together on this small world far better than hitherto, or is this far too optimistic?

What about making it easier for ETI to detect us? Should *we* attempt interstellar communication?

14.8   COMMUNICATING WITH EXTRATERRESTRIAL INTELLIGENCE (CETI)

Deliberate attempts by us at interstellar communication are called Communication with ETI, or CETI (pronounced 'keh-tee'). An early proposal for CETI was in 1826, by the German mathematician, astronomer, and physicist Carl Friedrich Gauss. At the time it was thought that the Moon might be inhabited, and so Gauss suggested that we could signal to the 'Selenites' by cutting down areas of Siberian forest to illustrate Pythagoras's theorem for right-angled triangles, as in Figure 14.15. It was never attempted.

In the modern era the first major attempt at CETI started in 1972 with the launch of the spacecraft Pioneer 10. Under the inspiration of Frank Drake and the US astronomer Carl Sagan, it carried a metal plaque etched as in Figure 14.16. The man and woman between them show characteristics of all races, and their size is given by the drawing of Pioneer 10 behind them. Center left are lines drawn from the Solar System to various pulsars, with the pulsar period encoded along the line. This gives our position in the Galaxy. At the top is a diagram illustrating the 1,420 megahertz emission of hydrogen atoms (Section 14.3), and at the bottom is a diagram of the Solar System. Pioneer 10's main mission was a fly by of Jupiter, which it successfully achieved in 1973. At last contact, in January 2003, it was 81 AU from the Sun and will escape from the Solar System, though it is not heading toward any nearby stars. Pioneer 11, launched in 1973, carries a message with the same content. It too, after fly bys of Jupiter and Saturn, is heading out of the Solar System.

Voyagers 1 and 2, launched in 1977, also carry messages. In this case there is a copper disc with images and sounds of Earth. This was before the era of the CD,

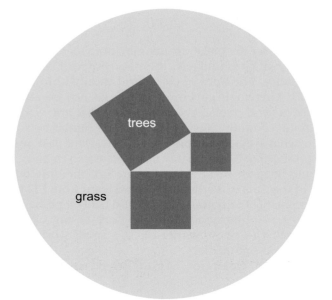

FIGURE 14.15 A way of signalling to ETI on the Moon through an illustration of Pythagoras's theorem.

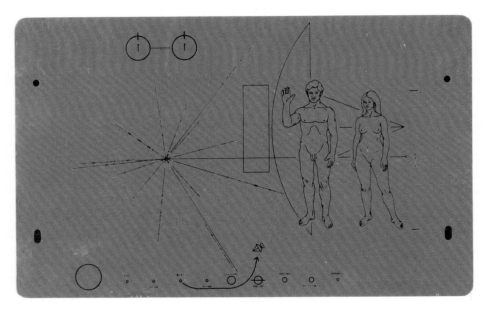

FIGURE 14.16 The message plaque on board Pioneer 10 and Pioneer 11, now leaving the Solar System. (NASA Ames Research Center, Lawrence Lasher)

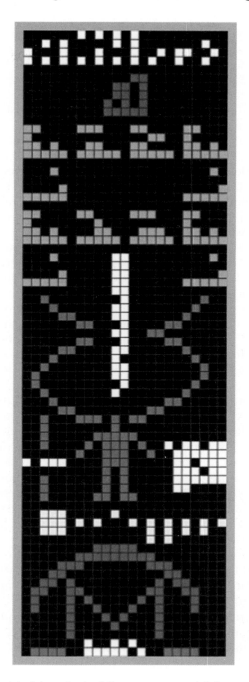

FIGURE 14.17 The pictorial content of the message sent into space in 1974 by the Arecibo radiotelescope. (Arne Nordmann)

so the disc is grooved and a stylus has to be run along it to recover the information. The container of the disc is etched with some of the information in Figure 14.16, plus diagrams illustrating how the disc has to be played. Voyager 1 flew by Jupiter and Saturn, and is heading for interstellar space, as is Voyager 2, which also visited Uranus and Neptune. As mentioned in Section 13.7, Voyager 1 is presently the furthest spacecraft from the Sun, at a distance of 101.4 AU – Neptune's orbit has a semimajor axis of just 30.2 AU.

 The probability of ETI coming across any of these spacecraft is extremely small, and even if ETI did find one, they might well not understand the messages. The first deliberate attempt at microwave communication at least had the merit of being targeted at stars. This was in 1974, when, at the opening ceremony of the refurbished Arecibo radiotelescope, the telescope was used to send a message to a globular cluster of about 300,000 stars called M13, about 21,000 light years away. The message will thus get there in about 21,000 years. Over the bandwidth, centered on 2,400 megahertz, the message outshines the Sun. It is in binary code, a total of 1,679 bits, and it took three minutes to transmit. Note that 1,679 is the product of two prime numbers, 23 and 73.

If correctly decoded the transmission shows a two dimensional blocky image that includes (among other things) an illustration of binary numbers, a simple drawing of the double helix of DNA, the outline of a human, and an indication that we live on the third planet from our star. You can probably make out some of this information in Figure 14.17. No location of the Sun is given, though this could be roughly determined by the direction from which the transmission came.

Given the large number of stars that will receive this short message in 21,000 years, there is some chance that we will have given away our existence far more effectively than by leakages of our radio, TV and radar transmissions. This was the basis of some protest about the Arecibo message at the time, though the low metallicity of the stars in globular clusters probably makes rocky-iron planets rare, and the proximity of stars to each other in globular clusters might destabilize planetary orbits. Nevertheless, radioastronomers have not deliberately targeted any stars subsequently, and this has always been the case for laser signals. Also, except possibly for defense radar, we have probably not leaked anything detectable at stellar distances. It might seem silly to be cautious about giving away our existence, but even though the chances of contact are slight, the consequences could be huge, so it deserves some consideration. This disinclination to reveal our presence is another solution to the Fermi paradox, if this is behavior common to ETI (Section 14.7).

A related question is whether we should reply if we do pick up a signal from ETI. This is still a matter of debate, but I suspect we would find the temptation irresistible. The decision might depend on what we learn about ETI from the contact, and on the way that humanity reacts to the momentous discovery that it is not the only technological species in the Galaxy.

15

What might the aliens be like?

We are not short of attempts to show what alien life forms might look like. In science fiction we have an abundance of images; in books, magazines, television programs, films, and in the past couple of decades, computer games. Figure 15.1 shows one of literally thousands of examples. In the majority of cases the aliens are intelligent and leave much to be desired in their behavior.

FIGURE 15.1 Martians, as depicted in an early edition of *The War of the Worlds*, by H G Wells, which was first published in 1898. This drawing is by Warwick Goble.

But do we have to rely on flights of fancy, or can science guide us toward more likely images? Let's consider alien biologies, first at the biochemical level, and then at the level of the whole organism, particularly large organisms comparable in size to large plants and large animals on Earth. At the top of the tree of life, what might ETI look like?

15.1   ALTERNATIVE BIOCHEMISTRIES

"It's life Jim, but not as we know it." So, how do we recognize life? What is life? Perhaps the least *un*satisfactory definition is along the following lines:

- life is a self-sustaining chemical system that undergoes evolution at the molecular level. In other words, it is a self-replicating, evolving system.

There are many self-replicating systems, such as the growth of a crystal from another crystal, but they lack evolution, and so they never acquire the complexity that we see in living systems. We thus require a chemical system that can not only replicate itself, but do so imperfectly in such a way that evolution takes place.

Life requires a system that can store, read, and write very complex genetic information. The element carbon forms a huge range of very complex molecules, and therefore, from a chemical point of view, it is no surprise that carbon is the core element of life on Earth. It is also no surprise that water is the liquid of life – it is abundant and it can dissolve (and transport) a huge range of substances, including many carbon compounds. It also participates in biochemical reactions (Section 3.3). But are there any alternatives to life as seen on Earth?

One small step

In considering alternative biochemistries, let's first take a small step away from the biochemistry that underpins all life known to exist on Earth today. The particular small step I've chosen consists of the use of amino acids not found in any known terrestrial organisms, and likewise for the bases in DNA and RNA. (Alternatively, I could have chosen a step in which the chirality of biomolecules was reversed – Section 4.3.)

Recall that amino acids are the building blocks of proteins, which fulfil many and varied essential roles in living systems (Section 3.2). Of over 100 amino acids known, only about 20 are used in known terrestrial organisms. Recall also that four bases are used in DNA, adenine (A), cytosine (C), guanine (G), and thymine (T), which can only pair as A-T and C-G. In RNA thymine is replaced by uracyil (U) (Section 3.2). Yet there are at least 12 bases that can be arranged in 6 mutually exclusive pairs. Could there be life using different amino acids, different bases?

There seems to be no reason why not. Indeed it is possible that a so called shadow biosphere exists on Earth today using different amino acids or bases! But this would not necessarily be alien – it could be of thoroughly terrestrial origin. What are the arguments for and against?

The great majority of biologists believe that, whereas there were probably multiple origins of life, only one form has survived. It is argued that other forms lost out in two ways.

- The various protocells exchanged genetic information resulting in just one type of protocell that led on to life as we know it.
- Even if biochemically different cells came into existence, only the most robust and aggressive type would have come through the competition for resources.

The few biologists who countenance the possibility of a shadow biosphere disagree. They counter the first argument by pointing out that geographical isolation of small regions of the Earth was likely, with consequent regional variations in the collections of organic molecules present. As a result, global homogenization of protocells was unlikely. Such isolation can be seen today. They counter the second argument by pointing out that if cell biochemistries were different enough, then the most aggressive type of microbe would have found the other types, and their food, unappetizing. Also, they claim, there could easily have been ecological niches, such as extremely dry deserts, that have been far less hospitable to our form of life than to microbes constituting a shadow biosphere.

It is thus concluded by a few biologists that microbes with different amino acids and/or different bases (and perhaps different in some other biochemical detail) could be present today living alongside our form of life. But if so, why has not a single example been detected? Figure 15.2 shows some cells, and colonies of cells. But are we really certain that their biochemistries are essentially the same? External appearance is no guide.

There is a lesson here from the bacteria, as you saw in Section 5.1. Until the 1970s all prokaryotes were placed in the kingdom Bacteria. Then, it was noticed that there were significant biochemical differences between certain bacteria and the rest. This led to the splitting of the prokaryotes into Eubacteria (usually called Bacteria) and Archaea. These are two of the three domains of life as we know it, the third being Eukarya, comprising the old kingdoms of Animals, Plants, Fungi, and Protoctista. We now know that there are greater genetic and biochemical differences between Bacteria and Archaea than between Archaea and Eukarya.

This was a major revolution in biology, and yet the evidence had been there for some decades, its significance unrecognized, so certain were biologists that there was no great biochemical distance between the various types of prokaryotic cell.

It is important for you to realize that in spite of the biochemical differences that distinguish the Bacteria from the Archaea, and the additional differences in cell structure that distinguish the Eukarya from the Archaea, the fundamental biochemistry is the same – the same amino acids, the same bases. We have not yet encountered a shadow biosphere. The Bacteria have been used as an example of how it is possible to overlook a major life form on Earth.

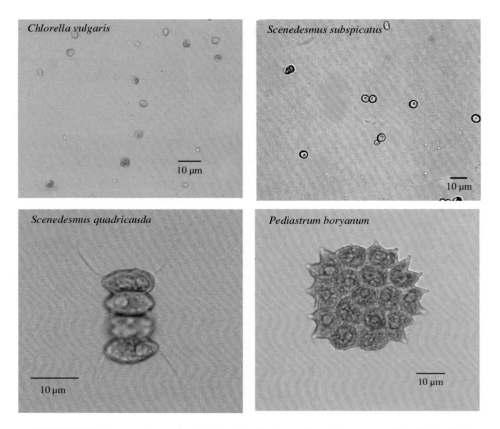

FIGURE 15.2 Some cells and colonies of cells. A μm is a micrometer. (Derek Martin)

It is certainly the case that a similar appearance under the microscope does not imply a similar biochemistry. To see if a shadow biosphere exists we need to examine the genetic make up and the biochemistry of cells. This requires the growth of microbes in the laboratory. Unfortunately most microbes cannot be grown in the lab. As a consequence, it will be very difficult to identify a shadow biosphere with the techniques currently available.

A shadow biosphere could also be detected through any traces of biochemicals it left behind, that have no clear derivation from our form of life. However, it is difficult to rule out such an origin, or an origin in non-biological processes.

Presently, the strong belief that there is no shadow biosphere is perhaps blinding people to its presence.

What are the consequences for the remote detection of life on an exoplanet if the alien biosphere was an analog of the sort of shadow biosphere that we've been considering? Bearing in mind that such a biosphere would still consist of organisms based on complex carbon compounds and liquid water, there would be no consequence at all if oxygen and methane were generated such that we could detect these gases from afar. Otherwise, the best indication would be the detection of any pair of gases well away from chemical equilibrium.

Unfortunately, even if there were atmospheric evidence for a biosphere, we could probably not distinguish what particular form of carbon-liquid water life was present – this would very probably require the acquisition of samples from the planet, for detailed analysis.

Some larger steps

I'll now be less conservative and contemplate life forms that are much more different from our form of life than considered hitherto.

First, consider liquids other than (salty) water. The liquid that is readily available as a solvent of molecules will determine the biochemistry. The common solvents are made from some of the most abundant elements – hydrogen (H), oxygen (O), and nitrogen (N). Liquid water (H_2O) is the most abundant solvent in the cosmos. The next most abundant is liquid ammonia (NH_3).

Consider H_2O mixed with NH_3, such as is thought to be present in liquid form in three of the large satellites of Jupiter – Europa, Ganymede and Callisto, and quite possibly in Saturn's large satellite Titan (Figure 2.3). This has a lower freezing point than pure water, depending on the proportions in the mixture. The minimum is about $-100°C$, thereby extending the possible domain of life to considerably lower temperatures than on the Earth. Life that has evolved where NH_3 is a significant presence in water will surely exhibit differences from our form of life, though in what ways is unknown. It is, however, likely that it will be based on complex carbon compounds and that these will include close analogs of DNA, RNA, and proteins.

Moving into more exotic terrain, there is the possibility of life based on liquid NH_3 *in the absence of water*. Such organisms would live at even lower temperatures, though with water being the most abundant solvent in the cosmos, an absence of water is unlikely. Also unlikely is life based on solvents such as hydrogen sulphide (H_2S), phosphine (PH_3), hydrochloric acid (HCl), and formamide ($H_2N.CH$). For H_2S and PH_3 there is the severe problem of dissociation by UV radiation and by oxygen. All of them suffer from the drawback that they can dissolve a far smaller range of compounds than H_2O. Formamide is the best bet.

Far more exotic would be life in liquid nitrogen, N_2. At the Earth's surface, liquid N_2 freezes at $-196°C$ at the pressure at the Earth's surface. At different pressures the freezing temperature will be different, though not hugely so at the pressures found on, or near, the surface of planetary bodies. Are there any molecules that could form long complex chains that could dissolve in a liquid as cold as N_2? There are just a few. Most relevant in considering exotic life are the silanols.

The silanols are silicon compounds. Next to carbon, the element silicon (Si) is the most prodigious in the range of large, complex molecules that it can form, though it is a poor second. Nevertheless, it is the best option in liquid N_2. A small silanol molecule consists of a few tens of atoms, mainly carbon and hydrogen, but, crucially, with one or two atoms of oxygen and silicon. There can also be an atom of another element, such as nitrogen. The molecules can get very large, and do not necessarily have to consist of identical units joined together, and therefore a very large silanol molecule can carry a lot of information. Whether this could constitute the genetic blueprint of an organism is unknown, but the possibility of silanol-liquid nitrogen life, though remote, reminds us that we must not be parochial in our search for alien life.

In the Solar System, N_2 is present in Enceladus, the icy satellite of Saturn, and in Triton, by far the largest satellite of Neptune, but it is unlikely that there are significant quantities of *liquid* N_2, if indeed any at all. This need not be the case on suitably placed exoplanets or their satellites.

Though life so very different from our form of life is very unlikely, how could we best detect it? Short of visiting a planet, we must rely on spectroscopy to detect chemical anomalies in any atmosphere or at the surface that defy explanation through ordinary chemical processes.

Rocky life?

As well as being a crucial ingredient of silanols, silicon has been considered as an alternative to carbon at far higher temperatures than can be withstood by carbon-liquid water life. In Section 4.3 the possibility was outlined that life on Earth began in the form of clay minerals. These are made from silicates – compounds of silicon, oxygen, a metal, and water. Recall that it is their intricate physical forms that have been proposed as the genetic information carrier and not the clay molecules themselves. Recall also that in this model the clay-based life gradually gave way to our form of life.

A prominent role for clays in the origin of life on Earth is controversial. Nevertheless, could silicon itself form the basis of alien life without being taken over by carbon-liquid water life? For example, a very large molecule could be built by joining silicate units SiO_4 to form a long chain, perhaps with various metallic elements at the join. But this structure cannot carry anywhere near as much information as carbon molecules – it cannot get long enough. Moreover, such chains are too stable at our ambient temperatures (e.g., are not readable in the manner of DNA), a stability that extends to temperatures much higher than those found in the Earth's biosphere. The strong consensus is that silicon *cannot* form the basis of alien life, and the notion that silicon can extend upwards the range of temperatures at which life might be found (Figure 15.3) is thought to be erroneous.

15.2   LARGE ALIENS

We are probably on firmer ground in speculating about the external form of alien organisms than we are about alternative biochemistries. This might surprise you,

FIGURE 15.3 Silicon-based life, existing at high temperatures, is thought to be very unlikely. (Julian Baum, Take 27 Ltd.)

but it is because there is thought to be a universal principle of adaptation to the environment.

I shall be particularly concerned with large organisms, by which I mean organisms that we could readily see without a microscope. Alien microbes have external forms that are also adapted to their environment, but I think they do not excite as much curiosity as organisms that we could pat or shake hands with.

Adaptation

The principle of adaptation in a version relevant to this section states that a species will have sense organs and an external form that is adapted to its environment. This means that two similar environments, whether on the same planet, on two different planets in the same planetary system, or on two planets in different planetary systems, would have creatures that display similarities in their sense organs and external form. This principle enables us to make an educated guess about what aliens would look like in any environment that we care to specify, based on our experience with life on Earth. Let's briefly examine the effects of gravity.

If there were a rocky planet with the same mean density as the Earth, but with twice its radius (and therefore eight times its mass), the surface gravity would be twice that at the Earth's surface. The energy cost of climbing high against the force of gravity would result in squat plants and animals. Living in water would also help – the water would help bear the weight. An absence of flying creatures would be likely, as would be the presence of ground hugging mists. Eyes would presumably be adapted to the light of the star after its passage through the misty atmosphere.

If there were a rocky planet with the same mean density as the Earth, but now with half its radius (and therefore one eighth of its mass), the surface gravity would be half that at the Earth's surface. This is not that different from Mars, where the surface gravity on this small planet is 38% that on Earth, though crucially different in that I am envisioning an alien planet that has retained its atmosphere. Now the life forms would be gracile. There would be birds, and high fluffy clouds. Figure 15.4 shows a possible humanoid life form on such a planet. The low gravity has resulted in tall, slender creatures.

Adaptation by alien photosynthesizing organisms would include adaptation to the spectrum of the radiation that reaches them from their star. In the case of photosynthesis on Earth, red light is utilized because many red photons reach the Earth's surface, and blue light because each photon carries a lot of energy due to its short wavelength (Section 13.6), and can thus drive certain photochemical reactions that less energetic photons could not. Green photons fall between two stools – less abundant than red photons but each is not as energetic as a blue photon. Consequently, the absorption of green photons is weak – green light is reflected away, which is why plants on Earth look green.

On other planets the spectrum reaching the surface could be different, due to the combined effect of a star with a different surface temperature from the Sun, and absorption of certain wavelengths in the planet's atmosphere or oceans.

FIGURE 15.4 A possible humanoid life forms on a low gravity world. (Julian Baum, Take 27 Ltd.)

Detailed considerations indicate that photosynthesis could have evolved so that photosynthesizing organisms have little use for yellow light, or red light, and thus appear yellow or red respectively. It is thought unlikely that organisms would have little use for the energetic photons constituting blue light, in which case photosynthesizers that are blue would be rare.

Adaptation of sense organs and external appearance to the environment is seen widely on Earth, which lends support to our belief that we can make educated guesses about what alien life forms will look like, as we have made in Figure 15.4.

We can also see adaptation in other ways, such as in respiration. On Earth, insects and plants respire by diffusion through their exterior surfaces. In spite of having internal tubes, insects are limited in size by the distance over which diffusion is sufficiently copious. Large animals have specialized internal surfaces, notably lungs, connected to the outside. We expect to find these further adaptations among aliens once we have the opportunity to examine their internal organs, and yet further adaptations when we examine their behavior.

Convergent evolution

Adaptation is intimately connected with convergent evolution. This is when two species with ancestors in different parts of the tree of life (Section 5.1) have evolved to look or behave like each other to cope with some aspect of their environment. There are many examples of convergent evolution on Earth.

One example is humming birds and a moth called the hummingbird-hawkmoth (Figure 15.5). These have similar sizes, body shape, and feeding by

FIGURE 15.5 A hummingbird-hawkmoth. (James McGlasson)

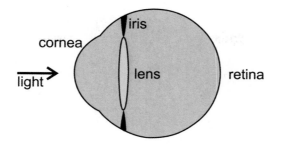

FIGURE 15.6 A camera eye, which humans and many other animals have.

hovering, yet the hawkmoth has shrimps for ancestors, and the hummingbird has dinosaurs!

Convergent evolution is also seen in body parts. Consider eyes. There are two main types. We, and many other species, see the world through two eyes, each of which resembles a photographic camera – camera eyes (Figure 15.6). Here, a single lens forms an image on the retina, which has many close packed light sensitive receptors, from each of which an electro-chemical signal is sent to the brain, providing a detailed view of the environment. This type of eye has evolved independently at least six times, each time through a long series of steps, presumably starting with a light sensitive patch. The adaptation is to an environment lit by the Sun – the retina is sensitive to the dominant wavelengths in solar radiation.

Two eyes give us binocular vision if they both point forward, as in humans and many other animals, including birds of prey. This has the advantage of being able to perceive relative distances at the sort of ranges that food lies, in the form of plants or animals. If the eyes are at the side of the head, as with many animals that are preyed upon, this provides a wide field of view, which facilitates the spotting of predators.

These adaptations have resulted in the evolutionary convergence of camera eyes in a wide variety of animals – octopuses, fishes, snakes, birds, mice, dogs, apes, humans, and many more.

The other main type of eye is the compound eye (Figure 15.7). This is the eye found, for example, in insects. Whereas the camera eye has a single lens forming an image on many receptors, the compound eye has many lenses, each one feeding light to a single receptor. Each lens points in a different direction, and so, as with the camera eye, an image is built up from the signals from the receptors. The compound eye needs to be far larger to achieve the same detail as a camera eye provides, because the sensors are spaced further apart – if we had compound eyes they would need to be about a meter across to give us the visual acuity our camera eyes give us. The compound eye does, however, have the advantages of being very sensitive to fast motion in the image, and of having a large field of view. It has evolved independently at least four times.

There are many other terrestrial examples of convergent evolution. Among

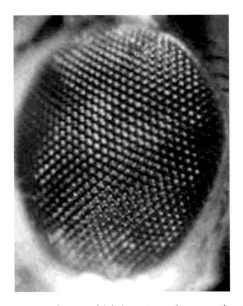

FIGURE 15.7 The compound eye, which insects and some other types of animals have. (Drosophila Melanogaster, © Carolina Biological Supply Company, used by permission)

sense organs there is convergence in hearing, smell, touch, and the detection of infrared radiation (which we perform with our skins), but also in senses that we and many other animals do not possess, at least not to any significant extent, such as echo location and the detection of electric fields. Convergent evolution also occurs in communication. For example, vocalization has emerged many times, and in its most advanced form is found in birds, whales, dolphins, apes, and humans.

Convergent evolution also extends to locomotion – legs, fins, flippers etc., and to the ways that bodies work. An example of the latter is warm bloodedness, found today in mammals, birds, and some ocean dwelling fish – three well separated strands in the animal branch of the tree of life. Warm bloodedness confers the advantage of a constant body temperature against variations in environmental temperature, thus avoiding the inactivity in the cold that characterizes, for example, insects and reptiles. Inactivity makes the animal vulnerable to predation, and limits food gathering.

Convergent evolution, like adaptation, is regarded as a universal principle, guiding us in our predictions of what the aliens might be like, and how they might function. What can these principles tell us about intelligence among aliens?

15.3   INTELLIGENT ALIENS

To try to define intelligence is to enter a minefield. Not long ago two dozen prominent experts were asked to define intelligence – the outcome was two dozen somewhat different definitions! One of the less problematical definitions was signed by 52 intelligence researchers in 1994 . . .

> "a very general mental capability that, among other things, involves the ability to reason, plan, solve problems, think abstractly, comprehend complex ideas, learn quickly and learn from experience . . . it reflects a . . . capability for comprehending our surroundings – catching on, making sense of things, or figuring out what to do."

This definition is clearly aimed at human intelligence, though many of these attributes can be found in other species, enabling us to agree (presumably) that a dog is more intelligent than a jellyfish. It is possible that these attributes would also be possessed by some aliens.

A more specific form of intelligence is technological intelligence. As well as all the attributes in the above definition, an essential extra is the ability to manipulate the environment. It is this specific type of intelligence that I am concerned with here. Only a technologically intelligent species could construct devices for signaling their presence across interstellar space, for example by means of radiotelescopes or lasers, or betray their existence through extensive modifications of their planet or planetary system. So, what can we deduce about their likely attributes?

The principles of adaptation and convergent evolution lead us toward the following conclusions about technologically intelligent aliens.

- They will have a large information processing unit. This will probably, but not necessarily, be concentrated in one part of the body, in which case it would constitute a brain.
- They will have at least two eyes, for binocular vision, each with high visual acuity, located near the top of the body to afford the best view of the surroundings. It is quite possible that the brain will be close to the eyes to reduce the distance over which signals have to travel.
- They will have digits (fingers, or toes, or tentacles, etc.) that will enable them to manipulate materials. This is where dolphins lose out – their fingers and toes have become enclosed in flippers as a result of adaptation to an aquatic environment.
- They will have some means of locomotion, such as two or more legs, or wings, or snake-like scales, or, more exotically, gas jets.

Note that the information processing unit (the brain) need not resemble our brain structures. An example on Earth is provided by the contrast between the New Caledonian Crow and the great apes (gorillas, chimpanzees, orang-utans). Whereas these apes have similar brain structures (similar to ours), the crows have a substantially different brain structure, yet all of these creatures learn to use

FIGURE 15.8 A New Caledonian Crow using a tool to obtain food. (Alex Weir and Alex Kacelnik, University of Oxford)

tools. Each generation of crows learns how to make probes and hooks for getting bugs from holes in trees and other inaccessible places (Figure 15.8). The great apes similarly fashion probes for getting food. This indicates that it is rather unlikely that the brain of an intelligent alien (technological or otherwise) will have a structure that resembles ours. Note that tool use by the crows and the apes is another example of convergent evolution, this time to solve a feeding problem. Their tool use falls far short of human use. They are unable to signal their presence even across the small gulf of space that separates the Earth from the Moon.

Figure 15.9 shows a speculation on what technologically intelligent aliens could look like. It shows an alien with two eyes, presumably camera eyes, and presumably sensitive to the wavelength range of the radiation it receives from its star. The eyes are backed by what looks like a large brain case. It has tentacles to manipulate objects, and these could also facilitate locomotion. Those ovoid objects around it are its eggs, from which its descendants will emerge. No radiotelescope or laser is visible – perhaps it can't send interstellar signals, or doesn't want to, or perhaps such devices are just out of the picture!

Figure 15.10 is a much more conservative speculation. Though clearly alien, they do not look very different from us. They are a few thousand years behind us

FIGURE 15.9 A truly alien intelligence. (Julian Baum, Take 27 Ltd.)

in technology, but that's a drop in the ocean of galactic time. Already, they could be looking into their skies and wondering if we are here, just as, at this moment, we are wondering if they are out there.

15.4   THE END

That completes my account of the search for potential habitats beyond the Solar System, of how we can determine whether they are inhabited, and if so, by what sort of creatures.

I do hope that we are not alone. This seems to me to be *extremely* unlikely. The issue could be put beyond doubt tomorrow if aliens were to contact us by signals or by visitation, or it could remain an open question for centuries, perhaps for far longer, perhaps even after humans have become extinct or have evolved into another species. We will never discover that we *are* alone; we can only discover that we are *not*. I believe that we *will* discover extraterrestrial life within or beyond the Solar System before the end of the century, perhaps even within your lifetime.

FIGURE 15.10 Technologically intelligent aliens not very different from us in their external appearance. (Adolf Schaller, via Terence Dickinson)

Glossary

Adaptation (in biology) A species will have sense organs and an external form that is adapted to its environment. (There are broader definitions, but this narrower one is relevant to this book.)

Adaptive optics An optical system for partially correcting the effects of atmospheric turbulence on a telescopic image.

Amino acids The 20 or so building blocks of proteins.

Aphelion The point in the orbit of a body when it is furthest from the Sun.

Archaea One of the three domains of life, the other two being the Bacteria and the Eukarya. Bacteria and Archaea are prokaryotes and almost all are unicellular (single celled). Eukarya are eukaryotes, and can be unicellular or multicellular.

Arcmin One minute of arc – 1/60 degrees of arc.

Arcsec One second of arc – 1/3,600 degrees of arc.

Astrometry The accurate positional measurement of a celestial object such as a star.

Astronomical unit AU The semimajor axis of the Earth's orbit around the Sun. It has a value of 149.6 million km. It is a measure of the average distance of the Earth from the Sun.

Bacteria One of the three domains of life, the other two being the Archaea and the Eukarya. Bacteria and Archaea are prokaryotes and almost all are unicellular (single celled). Eukarya are eukaryotes, and can be unicellular or multicellular.

Binary code Information encoded in two states, which can be thought of as 0 and 1.

Biosphere On the Earth, or on any other planet, the assemblage of all things living and their remains.

Brown dwarf A luminous object, made largely of hydrogen and helium, with a mass between about 13 and 80 times the mass of Jupiter (Jupiter is 0.08 times the mass of the Sun). It is too massive to be a giant planet, but not massive enough to have a main sequence phase of core hydrogen fusion.

Carbon-liquid water life	Life based on complex carbon compounds and liquid water. All terrestrial life has this basis.
CETI	Communication with extraterrestrial intelligence.
Chemical element	A particular chemical element has a particular number of protons (positively charged subatomic particles) in the nucleus of its atoms. There are 92 naturally occurring elements: hydrogen has just one proton in its nucleus, helium has two, and so on, up to uranium with 92.
Chemosynthesis	The creation of an energy store in a cell using chemical reactions that do not involve photosynthesis.
Chirality	That property of a molecule whereby it cannot be superimposed on its mirror image.
Classical habitable zone (HZ)	That range of distances from a star within which the stellar radiation would maintain water in liquid form over at least a substantial fraction of the surface of a rocky planet.
Column mass	The mass per unit area of a surface, such as the atmospheric mass per unit area of a planet's surface.
Convergent evolution	Exemplified by two species with ancestors in different branches of the tree of life that have evolved to look or behave like each other, or have similar body parts, to cope with some aspect of their environment. (There are broader definitions, but this narrower one is relevant to this book.)
Coronography	In optical telescopes, the suppression of radiation from one object more than another, to enhance the visibility of the latter e.g., the suppression of the radiation from a star more than that from an orbiting planet. It is achieved by a complex arrangement of components in the telescope.
Diffraction limit	A fundamental optical limit to the performance of a telescope that determines how much fine detail is visible in the image.
Doppler spectroscopy	The measurement of radial motion by measuring Doppler shifts in the spectral lines of a moving object.
Drake equation	An expression that displays the factors that determine the number of civilizations in the Galaxy that are detectable by us today.
Dyson sphere	A hypothetical structure surrounding a star that

	enables a civilization to divert much of the stellar radiation for its own use.
Dwarf star	A star in the main sequence phase of its lifetime.
Earth-type planet	A rocky-iron planet having a mass between about 0.3 times the mass of the Earth, and several times Earth's mass. Earth-type planets are endowed with various proportions of volatiles.
Earth-twin	An Earth mass planet orbiting a solar-type star about the same age as the Sun, in its star's classical habitable zone.
Eccentricity	A measure of the departure of an ellipse from circular form, the larger the eccentricity e the greater the departure. For a circle, $e = 0$.
Effective temperature	The temperature of the photosphere of a star from where its radiation escapes to space.
Electromagnetic spectrum	1 A strip diagram of the different types of electromagnetic wave classified by wavelength e.g., ultraviolet, visible, infrared. 2 A display of the (relative) quantities of radiation present at different wavelengths.
Ellipse	A geometrical shape, like a circle viewed obliquely. The orbit of a planet around a star, or a satellite around a planet, are (very nearly) ellipses.
Embryos	Rocky or icy-rocky bodies, ranging from a few hundred to a few thousand km diameter. They were an important stage in the formation of the terrestrial planets.
ETI	Extraterrestrial intelligence.
Eukarya	One of the three domains of life, the other two being the Archaea and the Bacteria. Bacteria and Archaea are prokaryotes; Eukarya are eukaryotes.
Eukaryotic cell	More complicated than the prokaryotic cell, with internal structures called organelles. It came after the prokaryotic cell, and evolved from it. All multicellular creatures are made of eukaryotic cells.
Evolution by natural selection	Evolution of organisms though the occurrence of random mutations that prove to be beneficial to the survival and reproduction of descendants.
Exoplanetary system	A planetary system of a star other than the Sun.
Extremophiles	Organisms that live in what we would regard as extreme environmental conditions.
Fermi's paradox	This states that if extraterrestrial intelligence exists, then it must be widespread, in which case why aren't they among us?

Flare	A sharp, transient increase in the emission of radiation and charged particles from a restricted region of the photosphere of a star.
Galaxy, the	The assemblage of about two hundred thousand million stars (and interstellar matter) in which we live. Many of these stars we see as the Milky Way.
Giant planet	1 A planet consisting largely of hydrogen and helium, many multiples of ten times more massive than the Earth e.g., Jupiter and Saturn. 2 A planet (sometimes called a subgiant planet) a few tens times the mass of the Earth, consisting largely of icy and rocky materials, but with substantial quantities of hydrogen and helium e.g., Uranus and Neptune.
Giant star	The phase in the lifetime of a medium or low mass star that succeeds the main sequence phase. The upper limit of mass is about ten times the mass of the Sun.
Globular cluster	A cluster of hundreds of thousand stars, more tightly packed and much older than the stars that comprise open clusters. Globular clusters occur throughout the halo and disc of the Galaxy.
Gravitational microlensing	An apparent brightening of a background star produced by the gravity of a foreground (lensing) star.
Greenhouse effect	The name given to the phenomenon whereby the surface temperature of a planet is raised because the atmosphere absorbs little of the incoming stellar radiation but more of the radiation emitted by the planet's surface.
Habitable zone	See classical habitable zone.
Heavy bombardment	The bombardment of the inner Solar System by planetesimals, from about 4,600–3,900 million years ago. Bodies rich in water, carbon compounds, and other volatiles predominated.
Heavy elements	The chemical elements other than the two lightest elements, hydrogen and helium.
Hot Jupiter	A giant exoplanet in an orbit close to its star. For F, G, and K dwarfs the semimajor axis will be less than about 0.05 AU.
Hydrothermal vent	Place on an ocean floor where hot water and dissolved gases pour out.
Infrared radiation	Electromagnetic radiation with wavelengths longer than those of visible radiation, but shorter than those of microwaves.
Integration time	The time for which photons are accumulated.

Intelligence	Among many definitions: "a very general mental capability that, among other things, involves the ability to reason, plan, solve problems, think abstractly, comprehend complex ideas, learn quickly and learn from experience … it reflects a … capability for comprehending our surroundings – catching on, making sense of things, or figuring out what to do."
Interferometer	More than one telescope working together so that the detail in the image is much finer than each telescope alone could achieve.
Isotope	A chemical element with a specific number of neutrons (uncharged subatomic particles) in its nucleus.
Jupiter-twin	A Jupiter mass planet orbiting a solar-type, solar age star in an orbit with a semimajor axis of about 5.2 AU.
Kernel	Large planetary embryo, rich in water. Kernels could have been an essential stage in the formation of giant planets, by becoming massive enough to capture a large mass of hydrogen and helium from the nebular gas.
Last common ancestor	The most recent ancestor from which all life today on Earth originated.
Life (a definition)	Life is a self-sustaining chemical system that undergoes evolution at the molecular level.
Light curve	The variation with time in the quantity of light reflected by a body, such as a planet.
Light year	The distance that light travels in a vacuum in a year – near enough the distance travelled through space in a year. One light year is 9.460536 million million km.
Luminosity	The total power radiated by a body, over its whole surface, over all wavelengths.
Main sequence star	A star in that phase of its lifetime when thermonuclear fusion in its core is sustaining its energy output by converting hydrogen to helium.
M dwarf	Main sequence stars with the lowest effective temperatures. They have the lowest masses among main sequence stars, and very long main sequence lifetimes. They are the most abundant class of main sequence star.
Mass extinction	Relatively short period of time when a large proportion of species becomes extinct.
Megahertz	A wave going through a million cycles per second.

Metallicity	The proportion of heavy elements in a star i.e., elements other than hydrogen and helium.
Methanogen	Methanogens are chemosynthesizing prokaryotes in which methane appears as a by-product.
Microbe	A unicellular or multicellular creature too small to be seen without a microscope.
Micrometer	A millionth of a meter.
Microwaves	Electromagnetic radiation with wavelengths longer than that of infrared radiation, but shorter than that of a radio waves.
Mutation	Change in DNA through a variety of causes.
Neutron star	The remnant of a supergiant. A neutron star is about 10 km across, about the mass of the Sun, and made almost entirely of neutrons.
Nucleotide	A building block of RNA and DNA.
Nulling interferometry	An interferometer arranged so that radiation from a specific direction, such as a star in an exoplanetary system, is greatly suppressed.
Observational selection effects	Aspects of a technique for observing objects, in particular celestial objects, that favor the detection of certain types of object over others.
Open cluster	A cluster of a few hundred stars, less tightly packed and younger than the stars that comprise globular clusters. Open clusters occur in the disc of the Galaxy.
Opposition (of a planet)	As viewed from a planet, the configuration when a second planet lies in the opposite direction in the sky to the star at the center of the planetary system e.g., Mars as seen from Earth every 2.14 years on average.
Optical wavelengths	Optical wavelengths are electromagnetic waves with wavelengths in the ultraviolet, visible, and/or infrared parts of the electromagnetic spectrum.
Optical telescope	A telescope that works at optical wavelengths, invariably over some portion of the optical range.
Orbital period	The time that a body takes to go around its orbit once.
OSETI	The search for extraterrestrial intelligence at optical wavelengths.
Oxygenic photosynthesis	Photosynthesis in which oxygen is produced as a by-product.
Panspermia	The proposal that life on Earth originated from dormant cells that reached the Earth after journeying through space.

Perihelion	The point in the orbit of a body when it is closest to the Sun.
Photodissociation	The break up of a molecule (or atom) by the action of a photon.
Photometry	Measurement of the quantity of radiation received, for example, from a star.
Photons	A stream of particles that accounts for how electro-magnetic radiation interacts with matter.
Photosynthesis	The manufacture of organic compounds from inor-ganic compounds (carbon dioxide and water) by means of energy from photons.
Planetesimals	Rocky or icy-rocky bodies, 0.1–10 km across, which are potential building blocks of planets.
Plate tectonics	The process that is responsible for most of the re-fashioning of the Earth's surface. The rocky surface is divided into a few dozen plates in motion with respect to each other. They are created at some plate boundaries, slide past each other at other boundaries, and are destroyed at boundaries where one plate dives beneath another.
Prokaryotic cell	A relatively simple cell with no organelles. This type of cell predates the more complex eukaryotic cell.
Proteins	Complex carbon compound that play a variety of roles in cells. Some are structural, and others (as enzymes) catalyse (speed up) biochemical reactions.
Pulsar	A neutron star that we observe by the beam of electromagnetic radiation that it sweeps across us as it rotates, which we see as equally spaced pulses.
Radiometric dating	The dating of events in the history of a rock or a mineral by means of radioactive isotopes.
Radio waves	Electromagnetic radiation with wavelengths longer than those of microwaves.
Red edge	A sharp increase at near infrared wavelengths in the reflectance spectrum of the Earth due to green vegetation.
Resolution/resolving power	A measure of the fine detail that can be acquired by a telescope.
RNA world	A hypothetical stage in the emergence of life on Earth where RNA was used rather than DNA as the store of genetic information.
Self-replicating probe	A probe that, on reaching a planetary system would make more than one copy of itself, and send these out

	of the system. In this way the number of probes increases comparatively rapidly.
Semimajor axis	Half the longest dimension of an ellipse.
SETI	The search for extraterrestrial intelligence.
Shadow biosphere	A hypothetical second biosphere on Earth that differs slightly from the known biosphere at the biochemical level, for example, by having different amino acids in its proteins, and/or different bases in its RNA and DNA.
Silanols	Complex compounds based on silicon. It has been suggested that they could form the basis of life at extremely low temperatures.
Solar nebula	The circumsolar disc of gas and dust from which the planets grew.
Solar-type star	A main sequence star of spectra type G.
Species (life)	Two plants or two animals belong to different species if a fully developed male of one species and a fully developed female of the other cannot produce a fully fertile hybrid under natural conditions, or if its production is extremely rare.
Spectral class	For main sequence stars, the spectral class is determined by the star's effective temperature. The class is denoted by a letter. As effective temperature descends, the labels are O, B, A, F, G, K, and M.
Spectral resolution	The minimum interval of wavelength over which spectral features can be discriminated.
Speed of light	The speed with which electromagnetic waves (or photons) travel through space. Its value is 299 792.458 km per second.
Stromatolite	A structure comprising layers of minerals laid down by colonies of oxygenic photosynthetic bacteria – cyanobacteria. There are living examples and fossilized examples.
Supergiant star	The phase in the lifetime of a high mass star that succeeds the main sequence phase. The lower limit of mass is about ten times the mass of the Sun.
Tachyon	Hypothetical particle that can only travel faster than light.
Technological intelligence	As well as the attributes of intelligence (see above), the ability to manipulate the environment.
Terrestrial planets	Mercury, Venus, Earth, and Mars constitute the terrestrial planets i.e., the substantial rocky-iron bodies that dominate the inner Solar System. The term can be extended to include analogs of such planets in other planetary systems.

Tidal heating	The heating of a body, such as a planet or a planetary satellite, arising from the gravitational distortion caused by the planet's star or the satellite's planet respectively.
Transit photometry	The measurement in the decrease of the apparent brightness of a star when a planet in orbit around the star comes between us and the star.
Tree of life	A branching diagram that shows the relationship between the various forms of life on Earth.
Type I migration	A process whereby a planet migrates through a circumstellar disc without opening a gap in the disc.
Type II migration	A process that occurs at larger planetary mass than in Type I migration, after a gap has been opened in the circumstellar disc. The rate of migration is slower than in Type I.
Type I, II, III civilizations	Civilizations that utilize energy sources on increasingly large scales: Type I, planetary scale; Type II, stellar scale; Type III, galactic scale.
Ultraviolet (UV) radiation	Electromagnetic radiation with wavelengths shorter than those of visible radiation, but longer than those of X rays.
Warp drive	A hypothetical method of travelling faster than light, in which a spacecraft warps spacetime in its vicinity.

Further reading and other resources

BOOKS

This is by no means a comprehensive list, but is a representative selection. If a cloth bound edition is available then it alone is listed. Books significantly more advanced than this book are asterisked (*).

Astronomy (including instrumentation)

* Discovering the Solar System (2007), Barrie W Jones, 470 pages, Wiley, ISBN 978 0 470 01831 6
Discovering the Universe (2008), Neil F Comins and William J Kaufmann III, 550 pages approx., W H Freeman & Co., ISBN 978 1 429 20519 1
*The Sun and Stars (2004), Simon Green and Mark Jones (eds), 380 pages, The Open University and Cambridge University Press, ISBN 978 0 521 54622 5
*Galaxies and Cosmology (2004), Mark Jones and Robert Lambourne (eds), 448 pages, The Open University and Cambridge University Press, ISBN 778 0 521 54623 2
Foundations of Astronomy (2005), Michael A Seeds, 657 pages, Thompson Learning, ISBN 0 534 42128 8

Astrobiology (including life on Earth; some of these books include SETI)

Life in the Universe (2007), Jeffrey Bennett and Seth Shostak, 560 pages, Addison Wesley, ISBN 978 0 805 34753 1
* Planetary Systems and the Origin of Life (2007), Ralph Pudritz, Paul Higgs, and Jonathon Stone (eds), 328 pages, Cambridge University Press, ISBN 978 0 521 87548 6
Searching for Water in the Universe (2007), Thérèse Encrenaz, 193 pages, Springer-Praxis, ISBN 978 0 387 34174 3
Astrobiology: a Brief Introduction (2006), Kevin W Plaxco and Michael Gross, 272 pages, John Hopkins University Press, ISBN 978 0 801 88367 5
*Astrobiology (2005), Jonathan I Lunine, 452 pages, Pearson/Addison Wesley, ISBN 978 0 805 38042 2
The Living Universe: NASA and the Development of Astrobiology (2005), Steven

J Dick and James E Strick, 328 pages, Rutgers University Press, ISBN 978 0 813 53447 X

*Looking for Life: Searching the Solar System (2005), Paul Clancy, André Brack, Gerda Horneck, 364 pages, Cambridge University Press, ISBN 978 0 521 82450 7

*Life in the Solar System and Beyond (2004), Barrie W Jones, 317 pages, Springer-Praxis, ISBN 978 1 852 33101 6

*Life in the Universe: the Science of Astrobiology (2004), Iain Gilmour and Mark A Sephton (eds), 364 pages, The Open University and Cambridge University Press, ISBN 978 0 521 54621 8

Cosmic Company: The Search for Life in the Universe (2003), Seth Shostak and Alex Barnett, 168 pages, Cambridge University Press, ISBN 978 0 521 82233 6

The Origin of Life (2003), Paul Davies, 303 pages, Penguin, ISBN 0 141 01302 8

Earth, Life and the Universe (2001), Keith Triton, 240 pages, Curved Air Publications, ISBN 0 954 09910 9

Search for Life (2001), Monica Grady, 96 pages, The Natural History Museum, ISBN 0 565 09157 3

*The Search for Life in the Universe (2001), Donald Goldsmith and Tobias Owen, 580 pages, University Science Books, ISBN 1 891 38916 0

Rare Earth: Why Complex Life is Uncommon in the Universe (2000), Peter D Ward and Donald Brownlee, 367 pages, Copernicus Books, ISBN 0 387 95289 6

The Outer Reaches of Life (1995), John R Postgate, 290 pages, Cambridge University Press, ISBN 0 521 55873 5

Extraterrestrials: A Field Guide for Earthlings (1994), Terence Dickinson and Adolf Schaller, 64 pages, Camden House, ISBN 0 921 82087 9

Extrasolar planets

The New Worlds: Extrasolar Planets (2007), Fabienne Casoli and Thérèse Encrenaz, 197 pages, Springer-Praxis, ISBN 10: 0 387 44906 X

* Extrasolar Planets: Formation, Detection and Dynamics (2007), Rudolf Dvorak (ed.), 305 pages, Wiley, ISBN 978 3 527 40671 5

Distant Wanderers: The Search for Planets Beyond the Solar System (2001), Bruce Dorminey, 226 pages, Copernicus Books, ISBN 0 387 95074 5

Search for extraterrestrial intelligence (SETI)

Where is Everybody? (2002), Stephen Webb, 297 pages, Copernicus Books with Praxis, ISBN 0 387 95501 1

Is Anyone Out There? (1993), Frank Drake and Dava Sobel, 287 pages, Souvenir Press, ISBN 0 285 63138 1

ELECTRONIC MEDIA

There is a huge range of electronic media, and a high rate of updating and obsolescence. An annotated list of free astronomy software, websites, news, and much else, is available at the Sky Publishing Corporation website http://www.skypub.com/ It is updated monthly. I strongly recommend it.

WEBSITES

General

http://www.google.com/
   A very useful website for obtaining information on anything that can be expressed in no more than a few paragraphs.
http://wikipedia.org/
   An online encyclopaedia, with entries supplied by anyone. Generally OK, but the entries are not refereed.
http://www.astronomynow.com/
   The website of *Astronomy Now*, the leading UK popular astronomy monthly. Much information and news.

Astronomy societies

http://www.popastro.com/
   UK Society for Popular Astronomy (SPA). This society, with over 3,000 members, promotes astronomy as a hobby, with a particular focus on beginners, but catering also up to the highest levels.
http://www.britastro.org/baa/
   British Astronomical Association (BAA). It has about 3000 members and promotes astronomy as a hobby with a focus at rather higher levels than the SPA.
http://www.astrosociety.org/
   Astronomical Society of the Pacific (ASP). This caters for amateur astronomers at all levels.
http://www.planetary.org/
   The Planetary Society, which promotes and supports exploration of the Solar System.

Extrasolar planets (selection only)

Catalog by Jean Schneider
   http://www.exoplanet.eu/

Space agencies

European Southern Observatory
  http://www.eso.org/public/
European Space Agency
  http://www.esa.int/ http://sci.esa.int/
NASA
  http://www.nasa.gov/ http://www.jpl.nasa.gov/
Space Telescope Science Institute
  http://www.stsci.edu/

Telescopes (ground-based, selection only)

European Extremely Large Telescope
  http://www.eso.org/projects/e-elt/
Keck telescopes
  http://www.keckobservatory.org/
OGLE, gravitational microlensing survey
  http://bulge.astro.princeton.edu/~ogle/
SuperWASP, transit photometry survey
  http://www.superwasp.org/
VLT (Very Large Telescope)
  http://www.eso.org/projects/vlt/

Telescopes (space-based) and space missions (selection only)

Solar System exploration
  http://sse.jpl.nasa.gov/
Mars exploration
  http://mpfwww.jpl.nasa.gov/ http://mars.jpl.nasa.gov/
Cassini-Huygens mission to the Saturnian system
  http://ciclops.org/ http://sci.esa.int/huygens/
COROT, transit photometry survey
  http://smsc.cnes.fr/COROT/
Darwin, exoplanet spectra survey
  http://sci.esa.int/home/darwin/
GAIA, astrometry mission
  http://sci.esa.int/home/gaia http://www.rssd.esa.int/GAIA/
James Webb Space Telescope
  http://ngst.gsfc.nasa.gov
Kepler, transit photometry survey
  http://www.kepler.arc.nasa.gov/
Space Interferometry Mission
  http://sim.jpl.nasa.gov
Spitzer, infrared telescope

http://www.spitzer.caltech.edu/
Terrestrial Planet Finder (TPF)
   http://tpf.jpl.nasa.gov

Images

European Southern Observatory
   http://www.eso.org/
European Space Agency
   http://www.esrin.esa.it/
Hubble Space Telescope
   http://www.stsci.edu/resources
NASA
   http://photojournal.jpl.nasa.gov
   http://nssdc.gsfc.nasa.gov/planetary/

SETI

SETI Institute
   http://www.seti.org/
SETI@home screensaver, for participating in SETI
   http://setiathome.ssl.berkeley.edu/
SETI at optical wavelengths (OSETI)
   http://www.coseti.org

Index

Printing: Mercedes-Druck, Berlin
Binding: Stein+Lehmann, Berlin